AF819129

THE MYSTERY OF LIFE FOUND IN PLASMA

THE MYSTERY OF LIFE FOUND IN PLASMA

DR. DOUGLAS GABRIEL

Our Spirit, LLC
2025

OUR SPIRIT, LLC

P. O. Box 355
Northville, MI 48167

www.ourspirit.com
www.neoanthroposophy.com
www.gospelofsophia.com
www.eternalcurriculum.com

2025 Copyright © by Our Spirit, LLC

All rights reserved. No part of this publication may
Be reproduced, stored in a retrieval system, or transmitted,
in any form or by any means, electronic, mechanical,
recording, photocopying, or otherwise, without prior
written permission of the publisher.

ISBN: 978-1-963709-16-2

Contents

Plasma, the Fabric of Creation

"Modern humanity looks at the beginning of the earth, how it was formed from a primal plasma in space, how life, man, and—much like a mirage—the world of ideas arose out of this primal matter."

Rudolf Steiner, GA 214

Throughout history we have seen many reports of "fiery globes" entering churches during mass or visiting saints in their room after passing through a solid wall. In those days, they called them a "sign from God" or an angel or a divine visitation. Saint's halos or the radiance of a guru, all are light phenomena. Science calls these "light orbs" Unidentified Anomalous Phenomena after changing their name from Unidentified Flying Objects—science frequently changes the names of anything they don't understanding but want to play like they do. A recent "rash" of glowing "light orbs" has drawn the world's attention towards the skies and many are dumbfounded by the light show and don't know whether to attribute these "lights" to government anti-gravitation flying machines, space aliens, angels, climate change, the magnetic pole shift, Sun activity, or the End Time's prophecies, or perhaps Chicken Little was right, the sky is falling. Personally, I tend to choose the last opinion. All of the atmospheres of the Earth, out to the exosphere, have been polluted by satellites, space stations, bombarded by HAARP, filled with endless electromagnetic broadcasts, littered with space junk, attacked with aerial nuclear barium bombs, filled with metallic nanoparticles from chemtrails, and bludgeoned with experiments by mad scientists with little or no morality. Yes, the sky is falling, and it is not the Anunnaki aliens returning to claim the Earth they terraformed and then later seeded with human DNA.

In fact, those globes of light actually are the culprits that seeded Earth with oxygen (O2 and O3) and water—the principal elements necessary for life. Recent discoveries have proven what Dr. Rudolf Steiner had indicated a hundred years ago. Plasma in our plasmasphere and thermosphere (plasmoid cells to be exact) takes ionized helium from the Sun's plasma, and from interplanetary space, and mixes it with high density complex dusty plasma balls and transforms helium ions into organized oxygen (O3-ozone) and then later into the oxygen we find in air, water, and minerals that compose the solid earth. This magic transubstantiation happens in hot plasmoids which can magnetically contain temperatures above a million degrees Fahrenheit. These little "cells of creation" have continued to do this since the first atmosphere of Earth was formed. What is very interesting is that the same magnetosphere, plasmasphere, ionosphere, exosphere, mesosphere, stratosphere, and troposphere create and distribute ozone (O3) as a protective layer to absorb and deflect direct sun light and turn it into the ozone layer surrounding the Earth that allows life, including human, to come to birth. Before humans could incarnate, the ozone had to be in place. Earth's protective ozone shield was called by the ancients the "celestial mirror"; essentially, the ozone sphere and the plasmasphere are the wings of the Sun protecting her children on Earth.

Dr. Rudolf Steiner pointed out that the Moon helps these same protective atmospheric layers take ions of hydrogen (from Sun plasma and space dust particles) in a plasma form and brings them together in a plasmoid ball (cell) with hydrogen peroxide (H2O2) to form water molecules. Thus, it seems that our "loving" atmospheres had a plan to begin with when they started to create the basics of life on earth—oxygen and water.

In today's world, plasma is given every attribute, description, and name available, from: god, angels, demons, miracle substance, Sun containment field, break-through medicine, space propulsion, divine building blocks, unsolved mystery, non-atomic substance, foundation of all things in space, the mother of the elements, UFOs, UAPs, magic glowing balls of light, ball lightning, human consciousness, the bio-electric system of the body, and the life energy found in human blood. Strangely enough, plasmas are some of those things, but not necessarily the way people want to frame the picture. That is why we feel driven to do an expose on plasma highlighting its many forms and function. But, as you probably know, we are big supporters of a Theory of Everything that includes all seven ethers and plasma; not the nebulous "ether" that scientism, philosophers, and weak-minded thinkers use as an excuse for ignorance. There is nothing wrong with not knowing. We don't need to measure, number, and weigh every natural phenomenon and name it after a scientist who believes he is the first to "discover" it. Plasma is all about not knowing, mystery, primal creation, and realms where humans cannot tread. We can poke Mother Nature, but she doesn't appreciate being destroyed just to understand her.

Plasma defies all of the standard laws of physics, including: gravity, the weak and strong nuclear force, electricity, and any other standard scientific approach to plasma as an "atomic substance." Plasma *is* and *is not* a substance, and it seems to obey its own set of super-natural laws that scientism has yet to understand. And yet, like every "force" scientists have come across, they have used that force to make weapons and medical procedures that are untested, uncontrolled, often deceptive, and seldom considerate of the health and well-being of humans or the Earth. Plasma is over 99% of space substance, according to scientism. Yet, scientists take what they think they know about the 1% and try to apply it to the 99%. This is pre-school sandbox behavior. But then again, humanity is young, and plasma is very, very old.

The greatest wish of science is to accelerate particles to near light speed and explode them against other particles in an attempt to find the "God-particle" that will add to scientism's existing imaginary particles, inside their imaginary atoms moving in an imaginary force field of electrons arounds protons in a pathetic attempt to subsume the "moment of creation." In other words, science wants to be God. This is part of the scientific fraud of inventing transient theories to make weapons instead of positive discoveries that benefit humanity. If science wanted to study the moment of creation, neither a cyclotron nor a magnetohydrodynamic plasma fusion containment of a "Sun on Earth" will help. Studying plasma in the human body, the Earth's atmospheres, the plasmasphere, the magnetosphere, and the magnetic reconnection of the Earth with the Sun through plasma magnetic ropes will reveal much more about the genesis of matter on Earth than current misdirected experiments. Basically, "mad-man scientists" are attempting to play God and use plasma to destroy, not create; exactly as they did with atomic knowledge that ushered in the nuclear age of "eminent death" at any given moment.

We need to study the life processes of the seven ethers (warmth, light, sound/chemistry, and life—unfallen sound ether, unfallen life ether and akasha ether) and its counterparts in plasmas, not the death processes of atomic nonsense projected onto nature and the entire universe. Ethers bring life, electricity in plasmas bring separation and recombination in the processes of proto-life creation, the precursors to human life on Earth.

Scientism (scientific materialism) defines electromagnetism as flowing electrons creating a magnetic field. What happened to the protons in this naïve equation? Where does the magnetic field come from and why? Why are electricity and magnetism usually together? Why is the majority of space full of static (non-moving) electricity, dust, and plasma dust clouds? And for that matter: What is so-called "physical matter" if plasma balls can pass through sheets of metal, or any substance for that matter, and remain coherent and undisturbed by the "solidness" of physical substance. Solid matter is the physical substance that the new religion of scientism tells us is the only truth we can know. This limited and ignorant view is called materialism, and it cannot explain the characteristics nor the behavior of plasmas, which are non-material. Plasma defies materialism and leads directly to the processes of life-affirming wisdom, not life-denying death.

Plasma needs her sisters the "Seven Ethers", most especially the sound/chemical ether and the life ether to do plasma's work in the physical realms, down here on this diaphanously thin lithosphere/troposphere we humans call Earth. For plasma and any one of the ethers to join together requires conditions provided by both the Sun's plasma "breath" (solar wind) and the Earth's atmospheres; especially the plasmasphere and thermosphere. The plasma from the Sun, in its etheric formative forces of warmth, light, sound/chemical, and life ether, comes down to the Earth through the solar breath magnetically reconnecting the Sun and Earth via magnetic transfer flux portals in the form of huge magnetic plasma flux ropes from the Sun that reconnect with the Earth every eight minutes bringing fresh Sun plasma directly into the plasmasphere of the Earth. This breath of Sun-plasma either stays in the plasmasphere for later use, and to protect the Earth, or comes into the magnetic poles through the auroral electrojet stream of plasma coming from the plasmasphere and flowing into the exosphere and thermosphere. Once this fresh breath of Sun-plasma enters into a high temperature thermosphere atmosphere, it can turn into complex dusty plasma balls (cells) igniting magnetic plasma containment, sometimes in excess of a million degrees, which transforms simple ions, protons, and molecules of inorganic and organic substance into life-substance that is tailor-made for Earth's troposphere and lithosphere (sky and earth). Creating the primal elements of oxygen (02) and water (H2O) are some of the goals of these "non-corporeal" entities whom many people say are "sentient" and "alive." One thing is for sure, without this atmospheric process, there would be no life on Earth. Whether plasmoids are conscious or not we cannot say for sure; but, we should respect and study the wondrous processes of plasmas and not destroy turn them into weapons.

Plasma can be visible and shining brilliantly, or completely invisible, or even exploding; its forms, shapes, and appearances are many and varied. Plasma close to Earth can appear above thunderstorms as red sprites, trolls, ball lightning. As plasmoids in our troposphere, they are thought to be UFOs which most seasoned airplane pilots and astronauts have witnessed. Just as quickly as these

UFOs appear, they can disappear after making impossible maneuvers like sudden unimaginable turns or flying off at speeds that are astoundingly fast—no jet plane has ever "caught up" with one at high speed. They can appear out of water, air, land, or sky and disappear at will, or just stay still for prolonged periods of time, appearing to be stars. Plasmoids are drawn to electromagnetic energy and seem to dislike and disrupt it. They fly through airplanes in flight that are filled with people, and no one is hurt. They are known to have interiors that make magnetic helical streams of opposing electricity that wind into the familiar forms of DNA and RNA. They gather "dust" from air born biological beings from the surface of the Earth and mix it with carbon and other ionized elements from meteoric dust in the atmosphere to fuse together all that is needed for inorganic and organic life. Then, plasmoids naturally create what scientists dream of: Sun-like fusion here on Earth. Scientists are spending endless research dollars on their little "fusion on Earth", "Sun in a bottle" experiments.

Fusion scientists also take lab-made plasma and spin magnetic cables inside of plasma balls into a torus of energy that helps create plasma fusion, sometimes precipitating substance and even biological life. We intend to demonstrate in this book, the natural fusion characteristics of plasmoids with empirical and spiritual scientific research which should help a clear thinker pause and wonder why we waste billions of dollars on fusion generators that cannot sustain a balanced plasmic fusion containment field necessary for a fusion generator.

Ethers are invisible but can be perceived with supersensible organs of perception by an awakened person who can witness the etheric formative forces working through the seven ethers, the elements, elemental beings, spiritual hierarchical beings, and the substances science calls "atomic." The atomic theory only works on the Earth; but, to understand the ethers or plasma, you have to be able to imagine realms that scientism declares do not exist. Ethers and plasmas have both eluded scientific mathematical logic, which does not apply to cosmic forces and processes. It takes consciousness to properly study a natural phenomenon that has been living for billions of years and then make "judgments" or "conclusions" about that phenomenon from a human perspective. Essentially, all knowledge changes over time and there is no such thing as what is called today, "settled science." Science knows little about the forces of the universe, but is wonderful at building mechanical devices that manipulate, utilize, and take advantage of those forces of nature. Fortunately, they cannot "get their hands on", or their "minds around", the nature, function, or genesis of ethers and plasmas.

In this book, we will tell you about many of the forms and applications of plasma that may surprise you. Plasma is now used in every industry, just as graphene oxide has been for the past decade. Both of these tools of scientism have no longitudinal studies that demonstrate they are safe. In fact, graphene oxide is considered a toxin to every part of the human body and yet is ubiquitous in our lives, delivered through water, air, food, clothing, medicine, and drugs, among many others. Plasma is becoming the rage of science like graphene oxide was, and it also is little understood. Just look at the many failed attempts to create a plasma fusion generator/reactor. Those billions of dollars could have been spent studying plasmas in their natural states, and more than likely would have been more enlightening. Even at this point in the innovation game, plasma is hailed as the way to heal wounds, kill cancer cells, extend life and restore teeth and eyes. There is hardly a field of study or industry that does not claim plasma as their new answer to all their questions.

Plasmas can be hot or cold, light or dark, manifest or unmanifest, self-organized or destructive, self-propagating or belligerent, and can communicate, eat, breath, multiply, hunt, graze, fly, and many other things that "defy scientific explanation." *If plasma was understood, materialism would die.* All of these things, and many more, are happening in laboratories currently throughout the world, and no one is regulating, monitoring, testing for safety, or considering the human being in all this furious work with plasmas. The human being's welfare is seldom part of the inventor's equation. The same thing happened with nuclear energy, graphene oxide, polyethelene glycol, mRNA, and many other 'wonders of science' that fly the false flag of "claiming to help" and "making life better."

What plasmas are, where they come from, and what they have to do with humanity's future are questions that will only be answered in the future, or when humankind can use science to study life not death. ***All plasma research should be regulated and halted until fusion scientists know what they are doing.*** At this point, they are simply guessing, just like CERN experiments with atomic explosions. Cold plasma "ray guns" and "dental" treatments should be stopped until longitudinal studies demonstrate the safety of these experimental procedures. ***All use of plasma weapons (see lists below) must end, space satellites grounded, and missions into our atmosphere curtailed until thoughtful people consider what these powerful forces of plasma-life are really all about and whether our attacks on the atmosphere are effecting the cosmic work of creation.*** The old scientific methods of capture, kill, and analyze the object of study does not work, it only leads to new weapons. Children should not play with matches, guns, nuclear weapons, ethers or plasmas. Since humanity is obviously not yet moral or responsible enough to handle the forces of nature, as their actions demonstrate, innovation in technology should take a back seat until virtuous humans can put humanity back into scientific innovation.

Most everything in the first part of this book is taken from major researchers and current discoveries in scientific journals and other articles. The second part gleans the wisdom of the past from the teachings of the ancients. The third part is drawn from Spiritual Science. The first part is an introduction and presents atomistic scientific information, much like a glossary, while the second and the third parts of the book are taken from ancient wisdom and Rudolf Steiner's interpretation of that wisdom added to the facts of scientism. The conclusion of the book is Spiritual Science and a mix of speculation that attempts to paint a picture of the future of plasma innovation and its relationship to the human being.

It is difficult to fully address plasma without addressing the ethers. For those of you not familiar with the ethers, we recommend our book: *The Eternal Ethers: A Theory of Everything*. Otherwise, you can access our many other books on Amazon that often address the ethers, plasma, magnetohydrodynamic generators, and other such topics.

It takes courage to study plasma and the ethers because the systems are complex, open-ended, comprehensive, not bound by time or space, and takes a mind that is connected coherently to the heart to reimagine the entire world of nature and the cosmos in light of spiritual scientific revelations. Without Dr. Rudolf Steiner's insight, from his life-long work called Anthroposophy (wisdom of humanity), it would be nearly impossible to fit all the pieces together to see a unified and sensible cosmology that embraces both plasma-electricity and the etheric formative forces. Once a living Moral Imagination is created concerning these topics, the beings of Moral Inspiration begin to pour into the awakened higher being of the seeker of knowledge and present images of what future humanity will become as we metamorphose through our spiritual development.

Humanity is not yet ready to understand the inorganic entities of the life-body of the Earth and its atmosphere. That was made perfectly clear recently when thousands of "light orbs" (UFOs-UAPs) descended from the thermosphere to balance out the attacks and damage the military/industrial corporate world has perpetrated on the atmosphere. Many people immediately proclaimed that plasma balls were angels and messengers from God or the divine. This quantum leap of faith is often the tendency of people influenced by luciferic forces that wish to draw humans into the upper atmospheres prematurely where they will meet their early demise. Others slip into recursive fear, thinking that the government is doing something evil, or that some devil has come up from hell. Strangely enough, both extreme vantage points are somewhat correct. The gates of heaven and hell have been opened and the soul and spirit of humanity is engaged in a battle, with the help of angels, to fight off the devils that are swarming all around us. That is why every time a powerful natural force is discovered, it is immediately made into a weapon—fear, greed, hatred, envy, and the other deadly sins take over the motivation of the scientist-inventor. These new discoveries reveal to us that nature can be friend or foe depending on who is perceiving her.

Since May 2024, the upper atmospheres of Earth have been depleted of ozone 60% by a proton CME from the Sun. This is to be expected since we are in the middle of the 25th solar cycle at the peak of the solar maximum. Also, the magnetic excursion of the Earth which mimics the Sun's magnetic polar reversal, which happened in November of 2024, makes the Earth more susceptible to the solar forcing of CMEs, solar flares, coronal holes, etc. It is because of these solar factors that we say the "sky is falling" and the "gates of heaven and hell have been opened." From below we see red plasma balls coming up out of the ocean and large bodies of water; from above we see the upper atmosphere ozone depleted in one solar event; we also see the ocean currents collapsing, the jet streams collapsing, the greatest number of volcanoes active at one moment in history releasing blue and green plasma balls; earthquakes at record levels releasing blue plasma balls; hundreds of satellites piercing the atmosphere and leaving space trash until they fall back to earth by the hundreds, and a thousand other "signs" that the "sky is falling" and heaven and hell have been opened. Is it any wonder that the SKY is reacting to huge solar events, the Earth is revolting and fuming, and human stupidity grows by the day?

We are not proposing that we worship plasma, as Robert Temple and others have suggested, but that we humbly study it in ourselves and in nature to find the secrets of creation instead of blowing something up, destroying our atmosphere, or inviting evil into innovation. Plasma is a natural process of creation, cleansing, and maintaining life on Earth. Plasma is a gift from our Sun that we have yet to be thankful for. Plasma shows us what happened on the 'days of creation', where life comes from, and what its natural processes are.

Atoms don't exist; they are an unproven theory. What does exist is our living Earth that works with our loving Mother/Father the Sun, who gives us life. Demeaning the Sun by using Earth-bound scientific theories to define her is an insult. Then to add insult to injury, scientists audaciously imagine that they can label the Sun with a number, a weight, and a measure—the Earth-bound scientific method. To image that humans can understand the Sun or the universe is laughable. First, we would have to appreciate the gifts of life coming from the Sun, and then we might, after thousands of years of study, begin to understand the plasma nature of the Sun and the gifts of the etheric formative forces in the cosmos, Earth, and the human being.

If plasma is a living being, as many new theorists are rushing to suggest, in a effort to find God somewhere, then plasma might be laughing at us, along with the Sun and the whole cosmos. But once we humble ourselves before the magnificence of creation, we will find that everything in the cosmos in also found in the human being. Or as the Hermetic philosophers were known to say: "As within, so without, as above, so below, as the universe, so the soul."

Radiant Matter or Energy?

Sir William Crookes indicated in his scientific paper of 1879, entitled: *On Radiant Matter,* that plasma is "radiant matter" that is substantially different than earthly physical matter. It is interesting to note that Crookes, the father of plasma, made his discoveries in the very year that the Archangel (Archai) Michael came into his regency to guide the world. To paraphrase Crookes' general estimation of plasma we can say: In studying plasma we seem to have within our grasp and obedient to our control indivisible particles which constitute the physical basis of the universe. We have seen that in some of its properties that plasma (radiant matter) seems material, while in other cases it assumes the character of radiant energy of electricity and magnetism. We are looking at the border lands of science where matter and force seem to merge into one another, a mysterious realm between the known [seen] and unknown [unseen] which create the foundations of life on Earth and in the Cosmos. The greatest scientific problems of the future will find their solution in this border land between the physical and spiritual nature of the universe, and even beyond; here, it seems to me, lie Ultimate Realities, subtle, far-reaching, and wonderful.

Consciousness, Plasma, and the Tree of Life

At this point and time, the same official science that proclaimed Plasma Cosmology a false theory, has pronounced that plasma is not alive, sentient, or conscious. The reason they believe their laboratory theories is because they can make many types of plasma in the lab. It is really quite ingenious the many uses they devise for hot and cold plasma. So, if they can turn it on and off, and control and repeat plasma processes, scientists are willing to go forward with innovation without much understanding of its true nature. And as we have said, modern science has spent more money on magnetic plasma containment and fusion generators than almost anything else in history. There is no comprehensive plasma theory that incorporates Earth plasma devices with astrophysical research in plasma. Plasma is the elephant in the room, and even though research shows that plasma balls react to speech, and even thought, they still ignore the elephant trumpeting loudly. The funny thing is that some of those scientists "rode in" on that elephant because they accept that 99% of the universe is plasma, but they do not know where it came from, what it is, or how it works.

One common mistake that is very prevalent is to say that plasma in Earth's atmosphere, plasmasphere, and Sun is a mechanism to record all that happens in the world and throughout time. Plasma theorists wish to make plasma into the akasha ether, the original ether that the other six ethers arose from. Robert Temple and others believe that plasmoids, the plasmasphere, and the Kordylewski Clouds in the triangulated libration points of the Moon's orbit, are in fact the famed Akashic Records, which are often called the Halls of Memory, and is the place where all that has happened is recorded and can be accessed by clairvoyants. These theorists speculate that plasma clouds of all types are made of cellular structures, crystalline substances, and helical streams of plasma filaments organizing into higher order structures that can hold memories and information of all types. Theorists believe they are memory clouds that are not only sentient but are also self-conscious beings which they "discovered" by making observations of plasma behavior *in situ*. They believe that all the normal processes of an organic self-conscious entity can be observed in the behaviors of different types of plasma. Plasma is, as we pointed out, the catch-all for all unfinished Theories of Everything, and these wild speculations are an example of materialism projecting its false perspective upon the world.

Plasma, in its various states, conditions, and locations around Earth provides mechanisms for the etheric formative forces called the "Tree of Life", or "unfallen" sound (chemical) and life ether. By unfallen we are referring to the ethers of sound (chemical) and life that have not penetrated into Earth's atmospheres and condensed into inorganic and organic life on Earth. One might say that the pure sound (chemical) and life ethers are slowly digested by the magnetosphere, plasmasphere, and the Earth's atmospheric layers that have turned inside out and taken on cosmic forces of levity (anti-gravitational) and natural electricity (anti-electromagnetic) which feed plasmas, plasmoids, and the Tree of Life functions which produce oxygen and water for life on Earth. Plasma and the unfallen ethers work together to create the fruit and leaves of the Tree of Life and the four rivers of life-giving water that arise at the base of the Tree of Life. These may seem like simple-minded analogies, but they are actually the true "fairy tale" that created human life on Earth.

The Earth, and these processes of plasmas and ethers that make up the Tree of Life, cannot be fully accessed in their natural conditions even though we have flown numerous satellites through these spheres around the Earth. Much of the current data concerning these "homes" of plasma are included in this article as a quick reference. But, without understanding the overall picture of the interwoven activities of the seven ethers (including the akasha ether) and the many types of plasma, it will be hard to derive a useful picture of what is going on in these very difficult to study realms. Just like in the Garden of Eden where the original Tree of Life was held back in paradise away from the reach of Adam and Eve who had violated the Tree of Knowledge of Good and Evil, the unfallen ethers (Tree of Life) and their plasma counterparts still hide their secrets of the food and drink of the gods—the water of life and the fruit of the Tree of Life. If this analogy is true, then a comprehensive understanding of the ethers and plasma should bring us eternal life, of sorts, or at least the ambrosia and nectar of the gods, the golden apples of Hesperides, the waters of the fountain of life. What might be the strangest discovery of our studies about plasma is that the analogies of all world religions, myths, and legends fit perfectly into a plasma-etheric formative forces cosmos. Often, fairy tales do come true.

The Weaponization of Universal Plasma

In this article, we intend to answer many questions about plasma and supply research and evidence to support our presentation. Much confusion whirls around the science, mythology, and religion of plasma and it behooves us to study the seeming "source of life", which some call the "incubator" of our planet, our solar system, galaxy, and universe. Wherever you look in your studies, you will find the ubiquitous "substance" of plasma, or something named plasma that might seem to be wildly different than the magic "substance" that has created almost everything in the universe that is outside of our warm and cozy Earth atmosphere. Plasma is like our mother who protects us from the deadly rays of the Sun, the solar system's source of our plasma. Plasma can pass through physical matter, like magic, without losing its coherence; it moves as fast as light and yet can be invisible; it can stretch millions of miles with spiraling helixes of electricity and magnetism in filaments and ion jets, and incubate solar wind and dust particles into Earth's air, water, and minerals; create layers of ozone and plasma around the Earth that protect us from harmful solar and cosmic radiation; move through the Earth's atmosphere in glowing plasmoids that defy all scientific laws "governing" matter; and, even be confused with UFOs, angels, gods, and aliens. Plasma is a most inexplicable force, entity, and phenomena that confounds the human mind and is becoming the fastest growing technical tool in all branches of medicine, fusion energy, and astrophysics. Thus, we might have more questions at the end of our study on plasma than when we began. Suffice it to say, plasma is a mystery that longs to be solved. Or, as the Egyptian Goddess Isis was known to say: "I am Isis, I am the past, the present, and the future; no one has lifted my veil and lived."

A few essential questions that come to mind concerning plasma are:

- Are UFOs and UAADs plasma balls in our atmosphere?
- Is "ball lightning" a type of plasma, or Ezekiel's fiery chariot, or a path of ascension?
- Is plasma god, or an angel, or higher spiritual beings?
- Can plasma be turned into weapons like guns and bombs?
- Are we currently using plasma weapons?

- Are plasmoids in the thermosphere and exosphere biological sentient beings?
- What is the current state of the art in plasma physics?
- Has plasma incubated our oxygen and water in the plasma around the Earth?
- Are plasmoids inorganic electrical phenomena or biological life?
- Is plasma an ether? Do combinations of any of the seven ethers make plasma?
- Has a plasmoid ball (ball lightning) ever been caught and examined?

The New Physics of Plasma

When I began studying plasma in 1971 by reading Ben Bova's book, *The Fourth State of Matter: Plasma Dynamics and Tomorrow's Technology*, I immediately became an enthusiastic believer in the seemingly spiritual nature of this magic substance, which is not a substance. I had never heard a word about plasma in my education or personal studies. Plasma was as hidden as the seven ethers and yet seemed as essential to the universe as breathing. I tried to find people to discuss this most wondrous substance with, but no one knew anything about it. In today's world, plasma is used in thousands of ways that, once again, I had heard nothing of until I renewed my studies of plasma. It is obvious to me now that plasma is "out of the closet" and being used in many new scientific "Theories of Everything" as well as every field of science. Just like ether, which became the go to explanation for modern scientists, plasma is now seen as everything from God and the angels, to UFOs and Unidentified Anomalous

Phenomena. In our study of plasma, we will try to demystify and glorify the true nature of plasma in all its wondrous workings.

The plasma universe theory posits that plasmas are the fundamental building blocks of the cosmos, playing a crucial role in the structure and behavior of the universe. From the formation of stars and galaxies to the dynamics of space weather, plasmas exhibit complex and interconnected behaviors influencing the evolution of cosmic structures, and thereby our very beingness. We might call plasma the nursery of organic and inorganic matter throughout the universe and on Earth.

To understand the genesis of our Sun and Earth we must understand plasma physics. Even considering the idea that the "supposed" emptiness of outer space (vacuum) brings to us a rich domain of plasmas and electromagnetic energies, plasma cannot be the mechanism that brings to birth the etheric formative forces that create matter because plasma is not matter. Plus, it is important to remember that plasma itself works with the akasha ether, which is the primal space in which the universe is contained and in which the ethers and plasmas work together. As Rudolf Steiner often pointed out, the universe is finite.

Plasma has been found in two huge plasma dust clouds next to the Moon that are nine times bigger than the Earth, and some scientists have taken to calling them Earth's second and third moons. In their article, *Kordylewski Dust Clouds: Could They Be Cosmic 'Superbrains?'* Robert Temple and Chandra Wickramasinghe propose that the Kordylewski Dust Clouds (KDCs), which are principally made of plasmas, may exhibit behaviors resembling those of highly complex living entities, hinting at the possibility of spontaneous evolution within these structures. The authors draw parallels between the characteristics of the KDCs and the characteristics usually attributed to spiritual beings, such as angels, divine messengers, or even God, suggesting that these cosmic structures may possess profound intelligence with the ability to communicate and interact with other entities in space.

Robert Temple's book, *A New Science of Heaven* delves into much deeper details concerning the KDCs, living plasmas in space, and the implications for the emergence of an entirely new paradigm. He highlights such ideas as the potential for plasma crystals inside dusty complex plasmas to spontaneously form helical structures resembling DNA, quoting many scientific papers:

> "Complex plasmas may naturally self-organize themselves into stable interacting helical structures that exhibit features normally attributed to organic living matter. The self-organization is based on physical mechanisms of plasma interactions. As a result, each helical string composed of solid microparticles [dust] is topologically and dynamically controlled by plasma fluxes leading to particle charging. These interacting complex structures exhibit thermodynamic and evolutionary features thought to be peculiar only to living matter. This new complex 'state of soft matter' in light of the autonomy, evolution, progenity, and autopoiesis demonstrates principles used to define life. It is concluded that complex organized plasma structures exhibit all the necessary properties to qualify them as candidates for inorganic living matter that may exist in space provided certain conditions allow them to evolve naturally."
>
> Source: Tsytovich, Morfill, et al, *From Plasma Crystals and Helical Structures towards Inorganic Living Matter*, in New Journal of Physics, Volume 9, 2007

NASA's imaging in the thermosphere (where plasmoids are often found) has revealed the presence of self-illuminated plasmas, which have been observed to display life-like behaviors and morphologies. These plasmas, with sizes estimated to be from one to eight kilometers in length or diameter, have been linked to UFO/UAP sightings, suggesting that these plasmas may account for many historical UFO observations. The life-like properties of these plasmas, including their ability to form cellular structures and engage in complex behaviors, have led researchers to speculate that they may represent a form of proto-life or non-biological life in space.

What is Plasma?

The plasmas we are interested in describing in this article are the electrically charged gases, or "ionized" gases that are found in nature as well as in numerous man-made devices. In addition to surrounding you in the universe, plasmas play key roles in bringing you sunlight outdoors during the daytime, street lights at night, fluorescent lights in your offices, computer chips in your computers, treated and coated surfaces of more important objects in your life than you might imagine (including hardened artificial joints you might have in your body and high temperature turbine blade coatings for the jets you fly in), flat panel displays for your TV, gas lasers, the welded joints that keep buildings from falling apart, and radio transmissions around the world. Among the important applications expected in the future for plasmas are the production of electrical energy from nuclear fusion, environmental cleanup, and spaceship propulsion for interplanetary flights.

Modern science tells us that plasma is one of the four states of matter, along with gases, liquids, and solids. It makes up 99.9% of the visible substance in the universe. Much of this is astrophysical plasma, the plasma that exists in outer space and that is contained within stars, including our Sun. Atomic theory tells us that in plasma, some of the electrons separate and become free from neutral atoms (atoms that have an equal number of protons and electrons and thus a neutral charge). The resulting free electrons make plasma different from the other states of matter, where the electrons are held close to nuclei. When the atoms in plasma separate from their negatively charged electrons, they no longer have a neutral electrical charge. Instead, the atoms become ions—positively charged particles. Therefore, plasma is an ionized state made up of positively charged ions (including protons) and negatively charged electrons.

There are several reasons why electrons in atoms can separate and form plasma. In laboratory experiments, scientists can blast the atoms with high-voltage electricity, lasers, or electromagnetic fields to form plasma. In space, plasma may form from high-energy photons, including gamma rays, striking atoms. Plasma can also form in space when gravity increases pressure so much that the pressure super-heats gases. The high temperatures cause atoms to collide with each other, resulting in the electrons separating from the atoms, creating plasma and the beginnings of a star. Unlike gas, plasma can conduct electricity. Also, in gases, all the particles behave in similar ways. In plasma, however, electrons and ions behave and interact in very complex ways, which creates waves and instabilities and makes plasma terribly unpredictable.

Plasmas come in several types. Most plasma in the universe is what researchers call high-temperature plasmas. In these high-temperature plasmas, temperatures can be more than 10,000 degrees Fahrenheit, and all the atoms can be fully ionized. Low-temperature plasmas are different. The atoms are only partially ionized, and they can be incredibly cold or even room temperature. Another unusual type of plasma is high-energy density plasma, which scientists create in laboratories to study their unusual properties. High Energy Density (HED) laboratory plasmas are perhaps the most extreme states of matter ever produced on Earth. HED plasmas have properties not found in normal plasmas under ordinary conditions. For example, matter in this state may simultaneously behave as a solid and a gas. In this state, materials that normally act as insulators for electrical charges instead become conductive substances. To create and study HED plasmas, scientists compress materials in solid or liquid form or bombard them with high energy particles or photons. High Energy Density plasmas are an exotic state of matter found in astrophysical events such as the birth of stars and brown dwarfs, in laboratory fusion experiments, and in nuclear weapons explosions.

Scientists generate HED plasmas using high-power petawatt-class lasers, which generate instantaneous power levels equivalent to the output of the entire U.S. electrical grid. HED plasmas are created in the National Ignition Facility, which is the most energetic laser in the world with two megajoules of light energy delivered in 16 nanoseconds.

Of course, atomic theories are just theories, and it seems to me that scientists are playing with natural forces that are more powerful than they have the virtue to understand. But we can obviously see the sub-natural direction materialistic science tends to go when first experimenting with life forces. Atoms are a false theory. The "monad" of material substance is more a copy of the living body of the Earth and its magnetosphere, plasmasphere, and multi-layered atmosphere than a theoretical atom. This "monad" is also found in the human body as it evolves over vast periods of time. Plasma is found in the Sun and in the human heart. Plasma processes circulate in human blood and plasma is found in the germ cells of all humans, the blood, nerve cells, and skin, among other areas.

Human Blood Plasma

Blood is a body fluid in the circulatory system of humans that delivers necessary substances such as glucose, oxygen, water, fats, proteins, and minerals to all cells and transport metabolic waste products away from those same cells. Blood is composed of blood cells suspended in blood plasma. Plasma, which constitutes 55% of blood fluid, is mostly water (92% by volume), and contains proteins, glucose, mineral ions, and hormones. The blood cells are mainly red blood cells (erythrocytes), white blood cells (leukocytes), and platelets (thrombocytes). The most abundant cells are red blood cells. These contain hemoglobin, which facilitates oxygen transport by reversibly binding to it, increasing its solubility. Humans have an adaptive immune system, based largely on white blood cells. White blood cells help to resist infections and parasites. Platelets are important in the clotting of blood.

Blood plasma is a light, amber-colored liquid component of blood in which blood cells are absent, but which contains proteins and other constituents of whole blood in suspension. It is the intravascular part of extracellular fluid (all body fluid outside cells). It is mostly water and plays a vital role in an intravascular osmotic effect that keeps electrolyte concentration balanced and protects the body from infection and other blood-related disorders.

In other words, blood plasma is not the plasma we are addressing but human blood plasma has many similar characteristics in a microcosmic fashion. We will hear later in this article about the magic of long-lived memory blood plasma cells that are created in utero and last a person's entire life. These cells are the memory-bank for the blood's immune system and for recognizing and fighting bad bacteria. This is quite similar to what hydrogen peroxide and ozone do in the atmosphere. Strangely enough, long-lived plasma memory cells carry inside of them white blood cells that have hydrogen peroxide in them to attack pathogens in the blood, subsequently releasing ozone.

Plasmoids

A plasmoid is a coherent structure of plasma and magnetic fields, often called plasma cells. Plasmoids have been proposed to explain natural phenomena such as ball lightning, magnetic bubbles in the magnetosphere, comet tails, solar wind, solar atmosphereics, and the heliospheric current sheet. Plasmoids produced in the laboratory include field-reversed configurations, spheromaks, and dense plasma. Plasma is often emitted in these laboratory experiments not as an amorphous blob, but in the form of a torus or a ball.

Plasmoids appear to be plasma cylinders elongated in the direction of a magnetic field. They possess a measurable magnetic moment, a measurable translational speed, a transverse electric field, and a measurable size. Plasmoids can interact with each other, seemingly by reflecting off one another. Their orbits can also be made to curve toward one another. Plasmoids can be made to spiral to a stop if projected into a gas at about 10–3 mm Hg pressure. They can also be made to smash each other into fragments. There is some scant evidence to support the hypothesis that they undergo fission and possess spin.

A plasmoid has an internal pressure stemming from both the gas pressure of the plasma and the magnetic pressure of the field. To maintain an approximately static plasmoid radius, this pressure must be balanced by an external confining pressure. In a field-free vacuum, for example, a plasmoid will rapidly expand and dissipate. Some plasmoids have been formed in discharges with local magnetic field strengths on the order of 16,000 Tesla.

A plasmoid in the thermosphere or exosphere refers to a self-contained, coherent structure of plasma held together by magnetic fields, essentially a localized region of highly charged particles that can exhibit complex behaviors within the Earth's upper atmosphere, often observed as glowing moving formations; these plasmoids are formed due to the intense solar radiation interacting with the atmosphere at these high altitudes, creating a dynamic environment where plasma can coalesce into distinct structures around crystals and dust particles they absorb. When this happens, they are called 'complex dusty plasmoids.'

The high energy radiation from the Sun ionizes the atmospheric particles in the thermosphere, creating plasma; under certain conditions, this plasma can then condense into plasmoids, influenced by magnetic field lines. Plasmoids can vary in size and shape, appearing as spherical, flat sheets, elongated spiral forms, and filament-like structures.

Some research suggests that plasmoids can exhibit complex behaviors, including movement, merging, splitting, and even seemingly coordinated interactions with each other, potentially due to the interplay of magnetic fields and plasma dynamics.

While not directly visible to the naked eye, plasmoids can be observed by specialized instruments on satellites and spacecraft, sometimes appearing as glowing formations. Studying plasmoids in the thermosphere and exosphere is important for understanding space weather phenomena, including disruptions to satellite communication and power grids caused by solar storms.

Characteristics of Plasma

Mainstream science, for the most part, looks on the universe as electrically neutral and purely mechanical; a place where the weak force of gravity holds sway. Plasma Cosmology, by contrast, acknowledges the electrodynamic nature of the universe. Because plasma remains electrically charged in space, it is influenced more by electromagnetic forces than gravity. In fact, space, once considered mostly empty, has been found to be alive with plasma. Vast flows of charged particles have been discovered spiraling hundreds-of-thousands of light years across interstellar space. The most familiar examples of electrical plasmas here on Earth are neon signs and lighting, television screens, and electrical arc welding machines. Some types of super-heated flames and atmospheric lightning are also forms of plasma. Plasma is an excellent conductor of electricity. Because of its free-flowing electrons, its conductive properties far surpass those of copper and gold. Due to its interaction with electromagnetism, plasmas display a complexity in structure far exceeding that of matter in gaseous, liquid, or solid states. It has a tendency to form into cellular and filamentary structures. These structures derive from the fact that a charged particle flow (or current) produces a ring of magnetic fields around itself, 'pinching' plasma into multi-filamentary strands, as can be seen in both cosmic and more localized scales.

Plasma Cosmology

Plasma cosmology is a non-standard cosmology whose central postulate is that the dynamics of ionized gases and plasmas play important, if not dominant, roles in the physics of the universe at interstellar and intergalactic scales. In contrast, the current observations and models of cosmologists and astrophysicists explain the formation, development, and evolution of large-scale structures as dominated by gravity (including its formulation in Albert Einstein's general theory of relativity).

The original form of the theory, the Alfvén–Klein Cosmology, was developed by Hannes Alfvén and Oskar Klein in the 1960s and 1970s, and posits that matter and antimatter exist in equal quantities at very large scales, that the universe is eternal rather than bounded in time by the Big Bang, and that the expansion of the observable universe is caused by annihilation between matter and antimatter rather than a mechanism like cosmic inflation.

Cosmologists and astrophysicists who have evaluated plasma cosmology reject it because it does not match the observations of astrophysical phenomena, as well as the currently accepted Big Bang model. Very few papers supporting plasma cosmology have appeared in the literature since the mid-1990s.

Alfvén–Klein's Plasma Cosmology

Hannes Alfvén suggested that scaling laboratory results can be extrapolated up to the scale of the universe. A scaling jump by a factor 10 to the 9th power was required to extrapolate to the magnetosphere, a second jump to extrapolate to galactic conditions, and a third jump to extrapolate to the Hubble distance.

In the 1960s, the theory behind plasma cosmology was introduced by Alfvén, a plasma expert who won the 1970 Nobel Prize in physics for his work on magnetohydrodynamics. He proposed the use of plasma scaling to extrapolate the results of laboratory experiments and plasma physics observations and scale them over many orders of magnitude up to the largest observable objects in the universe. In 1971, Oskar Klein, a Swedish theoretical physicist, extended the earlier proposals and developed the Alfvén–Klein model of the universe, or "metagalaxy", an earlier term used to refer to the empirically accessible part of the universe, rather than the entire universe including parts beyond our particle horizon.

In this model, the universe is made up of equal amounts of matter and antimatter with the boundaries between the regions of matter and antimatter being delineated by cosmic electromagnetic fields formed by double layers, thin regions comprising two parallel layers with opposite electrical charge. Interaction between these boundary regions would generate radiation, and this would form plasma. Alfvén introduced the term ambi-plasma for a plasma made up of matter and antimatter and the double layers are thus formed of ambi-plasma. According to Alfvén, such an ambi-plasma would be relatively long-lived as the component particles and antiparticles would be too hot and too low-density to annihilate each other rapidly. The double layers will act to repel clouds of opposite type, but combine clouds of the same type, creating ever-larger regions of matter and antimatter. The idea of ambi-plasma was developed further into the forms of heavy ambi-plasma (protons-antiprotons) and light ambi-plasma (electrons-positrons).

Plasma Applications

Some researchers believe that plasma could probably be the true sentient brain of the cell. Electromagnetism allows you to communicate with all cells, going beyond the mere border of the cytoskeleton and allowing adaptation; communication occurs simultaneously in the cell and between cells. Researchers have also created new types of plasma by combining ultrashort laser pulses and ultra-cold atomic gases.

Scientists have created HED plasmas by firing ultra-high-intensity lasers at titanium foil. These plasmas are so dense that they are almost solid, which allows physicists to explore new conditions.

Non-thermal plasmas are used to remove chemical and physical contaminants. They create radicals and reactive species, and also increase the temperature locally, emit UV light, and generate shockwaves.

Tokamaks are devices that use magnetic fields to confine plasma. Scientists at the Princeton Plasma Physics Laboratory are working on using tokamaks to harness the power of fusion.

Plasma cutting has advanced with new safety features and mobile apps that provide cartridge data.

Plasma treatment can be used to tailor surfaces for biomedical devices and implants. These surfaces can have improved biocompatibility and antibiofouling properties.

Plasma technology can be used to convert renewable energy into electrical power, which can then be used to create chemicals.

A partial list of the better-known forms of plasma:

- Cold, warm, and hot plasmas
- Ultra cold plasmas
- Ionized plasma
- Collisional plasmas
- Neutral plasmas
- Dense plasmas

- Magnetic plasmas and non-magnetic plasmas
- Complex plasmas
- Dusty plasmas and grain plasmas
- Colloidal plasmas
- Liquid plasmas
- Plasma crystals
- Active and passive plasmas
- Ideal and non-ideal plasmas
- High energy density plasmas

Plasma in Medicine

Imagine blowing a cold breeze of helium on a bleeding wound and watching the blood coagulate as the wound starts to heal at a faster rate than it normally would. Teams of scientists with expertise in plasma science, microbiology, biochemistry, and medicine are working on making something like that actually happen. They do not use helium exactly, but rather helium (or other gases) in an activated state called "plasma."

If a gas is heated or subjected to high electric fields, any electrons present are accelerated to high kinetic energies. When these electrons collide with the atoms and molecules of the background gas, they are able to knock off and free more electrons. The atoms/molecules missing one or more electrons are called "ions." The newly freed electrons are accelerated by the applied electric field and, in turn, enter into ionizing collisions with other atoms and molecules.

The mixture of neutral atoms and molecules, ions, and electrons is what is referred to as "plasma."

Plasmas range in temperature from very hot to very cold. Medical plasmas are in the perfect range—hot enough to produce the reactive species needed for effective treatment, but cold enough to leave tissue unharmed. In the last decade, research on the use of low temperature plasmas in medicine has intensified, and today plasmas are poised to dramatically affect healthcare. Researchers have discovered ways that plasmas can be applied directly to living tissues to deactivate pathogens; to stop bleeding without damaging healthy tissue; to disinfect wounds and accelerate wound healing; and to selectively kill some types of cancer cells.

Plasmas produce chemically reactive atoms and molecules such as hydroxyl (OH) and atomic oxygen (O) that can kill harmful bacteria through oxidation. This is important because oxidation of the lipids and proteins that constitute a cell's membrane can lead to the complete disruption of the membrane. Researchers discovered that bacteria cannot cope with the harsh environment created by plasmas; bacteria died in large numbers in a matter of minutes or even seconds, depending on the strength of the bacterial strain. Scientists have exploited this property of plasma to develop plasma devices that can be used to sterilize medical tools and instruments without the risk of damaging them.

Reactive plasma species sustained near room temperature can rapidly kill bacteria, viruses and fungi deposited on the surfaces of surgical instruments and medical devices, including those made of heat-sensitive polymers. These special properties of cold gas plasmas offer the prospect of protection from highly infectious agents such as prion, a structurally mis-shaped protein that is responsible for "mad cow disease," and which is resistant to all commercial decontamination procedures. Reactive plasma species appear to cause little or tolerable damage to living animal and plant tissues, while being capable of destroying bacterial cells. This is because bacterial and mammalian cells have very different responses to chemical and physical stresses, such as those experienced with cold gas plasma treatments. Consequently, skin fibroblast cells are found to remain viable under plasma conditions that are lethal to *E. coli* cells. This ability of plasma to destroy bacteria while leaving animal and plant tissue unharmed is key to the development of important plasma applications, including food decontamination, skin disinfection, and tumor reduction.

Similar to chemotherapies, cold plasma treatments appear to be able to induce programmed death (apoptosis) among cancer cells, arresting the rapid proliferation of cancerous cells but with little damage to living human tissues. Ultimately, the success of this approach will depend on finding the right recipes of reactive plasma species and delivering them effectively to the complex environment of diseased sites in the human body.

While plasma can destroy pathogens at sufficient doses, at a low dosage it can also accelerate the multiplication of cells, an important step in the wound healing process. The ability of plasma to kill bacteria cells and to accelerate the proliferation of specific healthy tissue cells, known as the "plasma kill/plasma heal" process, has led scientists to investigate the use of cold plasma for wound care. Although plasma-based technology for wound care is still in the research phase, preliminary tests show signs of successful treatments of some types of chronic wounds.

Another development is the use of cold plasma in dentistry. Recent work has shown that plasma can be an effective method of controlling oral biofilms. In a laboratory environment plasma has been shown to inactivate bacteria that cause tooth decay, and periodontal diseases. Plasma has also been applied to dentin, the calcified tissue structure underneath the tooth enamel, where it was found to reduce infection. Plasma could potentially remove the infected tissue in the tooth cavity, ultimately replacing the dentist drill. These recent developments indicate that not too far in the future, plasma-based devices could be available to dentists, allowing them to treat oral-borne diseases effectively, with little pain to their patients.

Magnetosphere

The Earth's magnetosphere is a region of space surrounding our globe in which charged particles are affected by the Earth's magnetic field. Simply put, it is created by the Earth's active interior dynamo creating dipole magnetic field lines. Farther out, field lines can be significantly distorted by the flow of electrically conducting plasma, as emitted from the Sun (i.e., the solar wind). Planets having active magnetospheres, like the Earth, are capable of mitigating or blocking the effects of solar radiation or cosmic radiation. Interactions of particles and atmospheres with magnetospheres are studied under the specialized scientific subjects of plasma physics, space physics, and astronomy.

Inside Earth's magnetosphere, "magnetic bubbles" are localized regions within Earth's magnetic field where the magnetic field strength is significantly enhanced, creating a bubble-like structure, often formed by the interaction of the solar wind with the magnetosphere, resulting in pockets of plasma trapped within the magnetic field lines; these bubbles can be dynamic and move around within the magnetosphere, particularly on the nightside of the magnetosphere's comet-like tail that is facing away from the sun. These bubbles are primarily formed when the solar wind, a stream of charged particles from the sun, interacts with Earth's magnetic

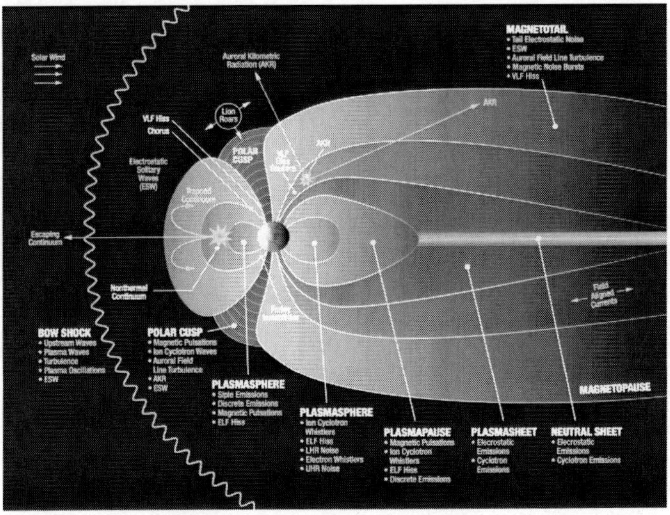

Earth's Plasmasphere

field, causing disturbances and compressions in certain areas. These active plasma balls are sometimes called "burst bubbles" because they stretch out field lines until they burst. These bubbles can trap charged particles, leading to variations in the intensity of auroras near the Earth's poles. Often these plasma bubbles are made primarily of highly charged protons (ions).

Kordylewski Clouds

Kordylewski clouds, also named ghost moons, are concentrations of dust that exist at the L4 and L5 Lagrangian points of the Earth–Moon system or the triangular libration points. They were first reported by Polish astronomer Kazimierz Kordylewski in the 1960s and confirmed to exist by the Royal Astronomical Society in October 2018.

The Kordylewski clouds are a very faint phenomenon, comparable to the brightness of the gegenschein, a faint oval-shaped glow in the night sky that appears directly opposite the Sun. They are very difficult to observe from Earth but may be visible to the unaided eye in an exceptionally dark and clear night sky. Most claimed observations have been made from deserts, at sea, or from mountains. The clouds appear somewhat redder than the gegenschein, indicating that they may be made of a different kind of particle.

The Earth does not pass through the Kordylewski clouds because they are not physically located in Earth's path; instead, they are faint dust clouds situated at the Earth-Moon Lagrange points (L4 and L5), meaning they orbit around a stable point in space where the gravitational forces of the Earth and Moon balance out, effectively staying in the same relative position to Earth and not crossing its path.

The confirmation in 2019 of the existence of the Kordylewski Clouds, and of their vast size which far exceeds the size of the Earth, suggests that the Earth-Moon System contains four, not two, major bodies, with the clouds dominant in terms of size. We point out that these clouds must be dusty complex plasma entities, which means that they will have an almost unimaginably rich interior structure consisting of regions of positive charge, negative charge, net zero charge, in gaseous, liquid, and crystalline states. Some people believe that capacities for high intelligence must have emerged in these clouds over the eons.

In two recent publications by Sliz-Balogh et al (2018, 2019) researchers focused their attention on the particular dust cloud (KDC) at the L5 point of the Earth-Moon system using sensitive polarimetric techniques. They found clear evidence that a cloud of submicron dust does indeed exist there through examination of polarized scattered light that varied with time. The cloud appeared to be "dynamic" and contained within it smaller dust clouds, perhaps displaying a cellular-like structure. Although evidence was cited to support the presence of ferric and silicate particles we cannot rule out, on the basis of available evidence, the presence of a dominant contribution of carbonaceous or organic grains, as indeed is known to be predominantly present in the interplanetary zodiacal cloud.

The temperature of Kordylewski clouds is estimated to be extremely cold, around -200°C (-328°F), as they are located at the Earth-Moon Lagrange points, which are regions of very low temperature due to their position in space and lack of direct sunlight exposure.

Earth's Plasmasphere

The plasmasphere, or inner magnetosphere, is a region of the Earth's magnetosphere consisting of low-energy (cool) plasma. It is located above the ionosphere. The outer boundary of the plasmasphere is known as the plasmapause, which is defined by an order of magnitude drop in plasma density. In 1963, American scientist Don Carpenter and Soviet astronomer Konstantin Gringauz proved the plasmapause's existence from the analysis of very low frequency (VLF) whistler wave data. Traditionally, the plasmasphere had been regarded as a well-behaved cold plasma with particle motion dominated entirely by the geomagnetic field and, hence, co-rotating with the Earth.

In 2014, satellite observations from the THEMIS mission have shown that density irregularities such as plumes or biteouts may form. It has also been shown that the plasmasphere does not always co-rotate with the Earth. The plasma of the magnetosphere has many different levels of temperature and concentration. The coldest magnetospheric plasma is most often found in the plasmasphere. However, plasma from the plasmasphere can be detected throughout the magnetosphere because it gets blown around by the Earth's electric and magnetic fields. Data gathered by the twin Van Allen Probes show that the plasmasphere also limits highly energetic ultra-relativistic electrons, from cosmic and solar origin, from reaching low earth orbits and the surface of the planet.

The Earth's plasmasphere is an inner part of the magnetosphere. It is a doughnut-shaped region of low energy charged particles (cold plasma) centered around the planet's equator and rotating along with it. It forms a plasma cloud that enshrouds the Earth. Its toroidal shape is determined by the magnetic field of Earth.

The plasmasphere is located above the ionosphere and extends outwards, with the outer boundary varying (depending on geomagnetic conditions). The size of the plasmasphere changes, depending on the solar wind–magnetosphere interaction.

The relations between the position of the plasmapause and the position of the radiation belt boundaries have been studied using the four Cluster satellites. For instance, during high geomagnetic activity time periods, the plasmapause is located closer to the inner boundary of the outer radiation belt.

Scientists have developed a 3D dynamic model of the plasmasphere that provides the number density of the particles and the position of the plasmapause as a function of time. Plasmaspheric density can also be determined from ground-based measurements, for instance from whistler electromagnetic waves recorded with a VLF (very low frequency) antenna.

The plasmasphere feeds the Earth's atmospheres via the auroral electrojet into the ionosphere.

The three panels below show how the relative locations of the outer boundary of the Earth's plasmasphere, the plasmapause, (shown in blue) and the Van Allen Belts (shown in red) change according to geomagnetic conditions.

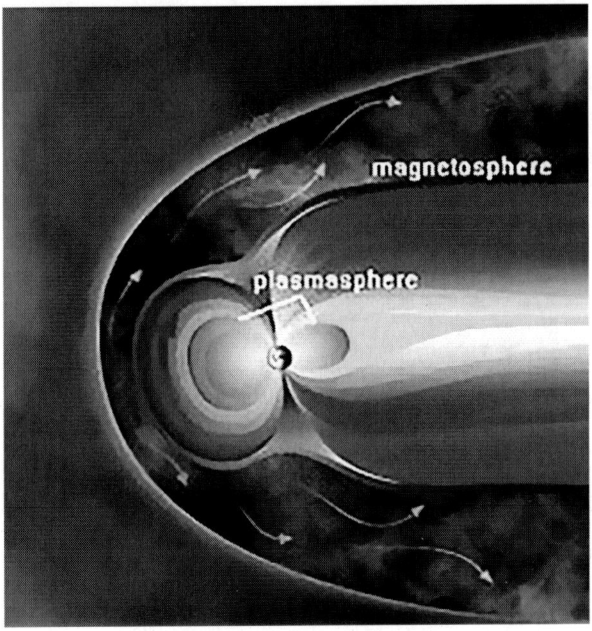

The Earth's plasmasphere and magnetosphere

The plasmasphere—the innermost part of the Earth's magnetosphere—is a doughnut-shaped region of low energy charged particles (cold plasma) centered around the planet's equator and rotating along with it. Its toroidal shape is determined by the magnetic field of Earth. The plasmasphere begins above the upper ionosphere and extends outwards, with the outer boundary varying (depending on geomagnetic conditions) from 4.5 Earth radii (RE) to 8 RE.

The two Van Allen radiation belts are concentric, tire-shaped belts of highly energetic (0.1–10 MeV) electrons and protons, which are trapped by the magnetic field and travel around the Earth. These radiation belts partly overlap with the plasmasphere. The inner Van Allen Belt is located typically between 6000 and 12 000 km (1-2 Earth radii)

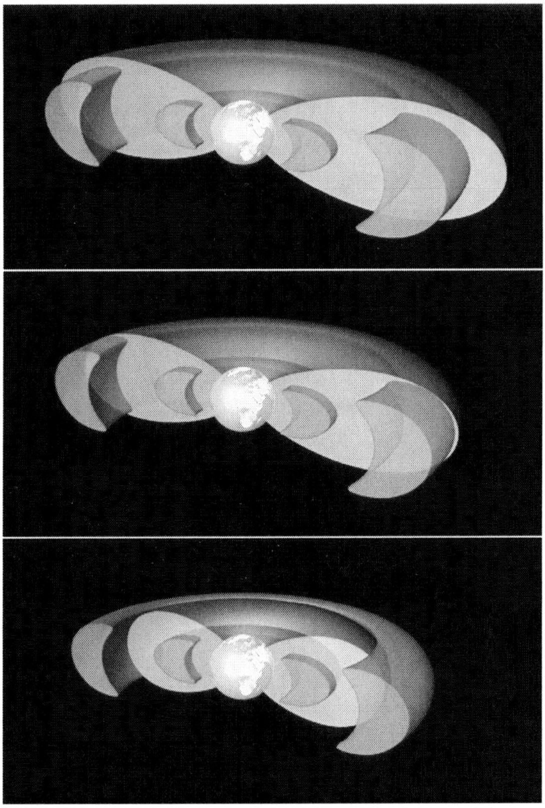

The plasmasphere and Van Allen belts

above Earth's surface, although it dips much closer over the South Atlantic Ocean. The outer radiation belt covers altitudes of approximately 25,000 to 45,000 km (4 to 7 RE). Both belts are separated from each other by an empty "slot" region.

Data from the Cluster mission have shown that the position of the plasmapause—the outer boundary of the plasmasphere—is quite variable and, in addition, that the size of the slot region between the radiation belts varies with changes in geomagnetic conditions.

During periods of low geomagnetic activity (top panel) the plasmapause typically extends to around 6 RE, occasionally expanding beyond the boundary of the outer radiation belt, as far as 8 RE or even further. This result contrasts with previous studies based on observations from different spacecraft which indicated a correlation between the position of the plasmapause and the location of the inner edge of the outer belt.

During periods of higher geomagnetic activity (with moderate activity illustrated in the central panel and high activity in the lower panel) the plasmapause moves closer to the inner boundary of the outer belt, to around 4.5 RE. This behavior is similar to that observed by previous studies. The size of the slot region also varies with geomagnetic conditions, being wider during low geomagnetic activity.

Since the May 2024 large proton CME struck the Earth (largest in 20 years), a new radiation belt of protons has been added to the existing belts. There were also two electron CMEs a days later that created two temporary electron belts that dissipated within a few weeks. The new proton radiation belt is unprecedented and is still in place.

Principles of Plasma and Higher Ethers

Plasma:

- A torus surrounds and embodies the vortex in a plasma cell (plasmoid) and helical filaments are swirled into toroidal nodes and shapes with great electrical and magnetic potentials.

- Filaments of plasma form as tubes, helices, cables, ropes, sheets, and can swirl into a torus.
- Gaseous cells of inorganic and organic materials incubating matter through fusion.
- Acts as an intelligent self-organized complexity with all the characteristics of a living entity.
- Can appear in lightning, ball lightning, tails of comets, stars, clouds, plasmoids, sprites, trolls, ion jets, relativity streams, and many other space phenomena.
- Possesses memory and forms helical structures like DNA that ostensibly hold data.
- Has double layers of + and—charges that create a "cell wall" structure that can magnetically hold electrical discharges of over a million degrees Fahrenheit.
- Double layers develop energy conversion and self-organization with plasma cells.
- A micro-fusion reactor inside a plasmoid cell that can change elementary particles into organized inorganic and organic molecules.
- Is stimulated by light to create oxygen from ionized helium in the exosphere and plasmasphere.
- Gaseous plasma cells can absorb inorganic and organic biological material and transform them into water and oxygen which helps create life on Earth.

- Electric skeletons for proto-cells in dust clouds that birth elementary Earth particles.
- An electron ensemble of electricity swirling into form.
- Electronic matrices guiding bioelectricity in living organisms.
- Displays complex space charge configurations.
- Are primary electrical biological cells creating prebiotic life for earth.
- Double layers create outer walls that capture and transform energy and feed off their environment.
- Manages the inflow and outflow of plasmoids from the plasmasphere into the exosphere and below.
- Demonstrates continuous synthesis of their environment from the outer magnetosphere down to the Troposphere.
- Plasma cells (plasmoids) breathe in fresh particles and organize them into higher order structures.
- Self-multiplies by division, mitosis, uniting with other plasmoids, and births other plasmoids.
- Communicates with other plasmoids through energy exchange.
- Turns thermal energy into electric current through its double layers.
- May be an inorganic double of the human being's bio-electrical energy.

- May be a purely electrical life-form.
- Provides the early building blocks of life-forms.
- Farms, grazes, and hunts in the thermosphere and exosphere.
- Cell matrices can form instantly with an electric arc of energy.
- Can form micro-bacteria in the laboratory when ball lightning explodes under water.
- Intertwined with humans in a symbiotic relationship.
- May be an impersonal energy not aware of itself consciously, working from electrical principles unknown to science.
- Emits infra-red with a heat signature, can glow brilliantly or be dark or invisible to the human eye.
- Exists in both the Earth's atmosphere and beyond in the magnetosphere, interplanetary space, and throughout interstellar space.
- Creates the Kordylewski Plasma Clouds that rotate around the Earth that have cell-like structures within and plasma filaments, torus shaped forms, magnetic ropes and cables, and cosmic dust in invisible clouds that are each three times the size of Earth.
- Gathers above hurricanes and thunderstorms and sends fantastic electrical charges high into the stratosphere and above.

Birkeland Currents

Birkeland currents are electrical currents that flow along the Earth's magnetic field lines, connecting the Earth's magnetosphere to the upper atmosphere (ionosphere-exosphere), essentially acting as a pathway for charged particles to travel between these regions; they are often associated with the formation of auroras and are named after Kristian Birkeland who theorized their existence based on his studies of the aurora borealis. These currents carry electrical energy from the magnetosphere towards the Earth's ionosphere, where they can interact with rising ionized atmospheric particles, causing auroral displays.

Birkeland currents are also referred to as "field-aligned currents" because they flow parallel to the Earth's magnetic field lines. They are generated by the interaction of the solar wind (charged particles from the Sun) with Earth's magnetosphere. Variations in solar activity can lead to fluctuations in Birkeland currents, potentially impacting communication systems and power grids. There are many types and sizes of these magnetic currents that arise from flowing electrical currents.

Ball Lightning

Ball lightning is a rare phenomenon that appears as a luminous sphere that can range in size. It can be transparent or translucent and appear in various colors, including red, orange, yellow, white, or blue. It is often described as moving erratically, sometimes slowly floating through the air or even passing through solid objects. Ball lightning can appear at ground level or be reported by airplanes and astronauts and can last for many seconds or a millisecond. It usually occurs

Ball lightning

near the ground during thunderstorms in close association with cloud-to-ground lightning.

Ball lightning is often accompanied by a hissing sound and distinct odor. Some accounts also mention an acrid smell, possibly from ozone or burning materials. The first modern recorded sighting of ball lightning occurred in 1638, when a "great ball of fire" came through the window of an English church.

There's no one-size-fits-all explanation for the properties of ball lightning. Some scientists suggest that ball lightning consists of mini black holes formed during the Big aBng or charged particles interacting with the atmosphere, or some other scientific fantasy.

Plasmoids, ball lightning, complex dusty plasmoids, and many other forms of plasma have been witnessed many times by pilots, astronauts, and scientists. I personally have had numerous encounters with ball lightning and can attest to its existence and the many strange phenomena that surround its appearance.

Red Sprites

Only relatively recently did the upper part of great thunderstorms come to light. Modern science and high-speed cameras began catching sight of amazing phenomena call "red sprits." Red sprites are considered a form of "cold plasma," meaning they are electrical discharges occurring in the upper atmosphere above thunderstorms, consisting of ionized gas that is not as hot as regular lightning, but still exhibits plasma-like characteristics; essentially, they are a luminous discharge within the atmosphere's plasma state. Sprites are classified as a type of "Transient Luminous Event." Unlike typical lightning, which is considered "hot plasma," red sprites are considered "cold plasma" due to their lower temperature and similarity to a fluorescent tube discharge.

When a strong positive lightning strike occurs from a thunderstorm cloud to the ground, it can trigger a large electrical discharge in the upper atmosphere, creating the plasma phenomenon of a red sprite. The red color of sprites comes from the excitation of nitrogen molecules in the low-pressure environment of the upper atmosphere.

TROLLs (transient red optical luminous lineaments) occur after strong sprites, and appear as red spots with faint tails, and on higher-speed cameras, appear as a rapid series of events, starting as a red glow that forms after a sprite tendril that later produces a red streak downward from itself. They are similar to jets.

Upper-atmospheric lightning and ionospheric lightning are terms sometimes used by researchers to refer to a family of short-lived electrical-breakdown phenomena that occur well above the altitudes of normal lightning and storm clouds. Upper-atmospheric lightning is believed to be electrically induced forms of luminous plasma. The preferred usage is transient luminous event (TLE), because the various types

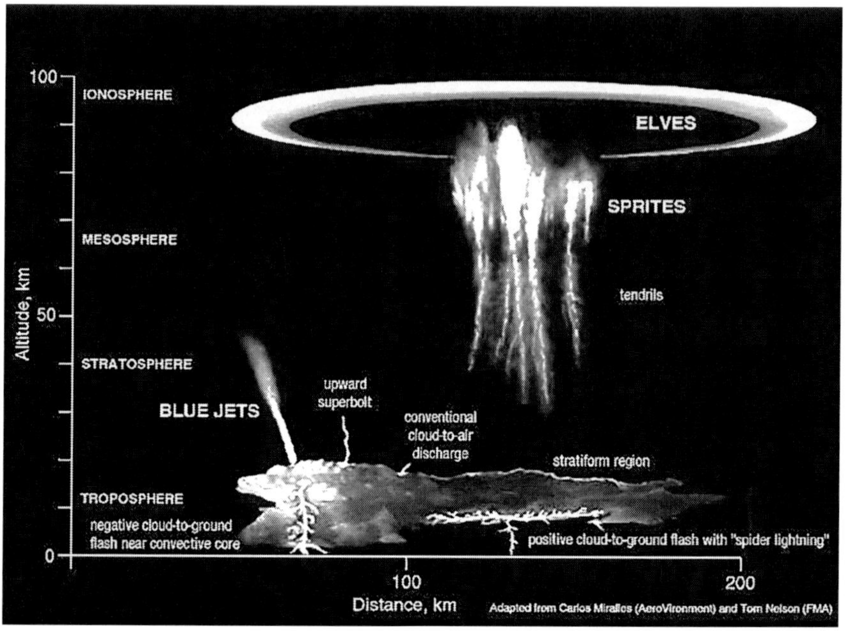

Upper-atmospheric lightning and electrical-discharge phenomena

of electrical-discharge phenomena in the upper atmosphere lack several characteristics of the more familiar tropospheric lightning.

There are several types of TLEs, the most common being sprites. Sprites are flashes of bright red light that occur above storm systems. C-sprites (short for "columniform sprites") is the name given to vertical columns of red light. C-sprites exhibiting tendrils are sometimes called "carrot sprites." Other types of TLEs include sprite halos, ghosts, blue jets, gigantic jets, pixies, gnomes, trolls, blue starters, and ELVES. The global rate of TLE occurrence has been estimated from satellite (FORMOSAT-2) observations to be several million events per year. Although jets are considered to be a type of upper-atmospheric lightning, it has been found that they are components of tropospheric lightning and a type of cloud-to-air discharge that initiates within a thunderstorm and travels upwards. In contrast, other types of TLEs are not electrically connected with tropospheric lightning—despite being triggered by it.

Blue jets are believed to be initiated as "normal" lightning discharges between the upper positive charge region in a thundercloud and a negative "screening layer" present above this charge region. Blue jets occur much less frequently than sprites.

Giant blue jet

ELVES often appear as a dim, flattened, expanding glow around 250 miles in diameter that lasts for just one millisecond. They occur in the ionosphere 62 + miles above the ground over thunderstorms. Their color was unknown for some time but is now known to be red. ELVES were first recorded on a shuttle mission. ELVES is an acronym for emissions of light and very low frequency perturbations due to electromagnetic pulse sources. This refers to the process by which the light is generated; the excitation of nitrogen molecules due to electron collisions; the electrons possibly having been energized by the electromagnetic pulse caused by a discharge from an underlying thunderstorm.

A picture of ELVES from Possagno, Italy

Plasma Jets in Space

A "plasma jet" in space refers to a powerful stream of electrically charged particles (plasma) that are ejected from a celestial object, most commonly a supermassive black hole, at extremely high speeds, often appearing as elongated beams extending far beyond the source galaxy. These jets can be incredibly large, spanning millions of light-years and are formed by complex interactions between plasma and strong magnetic fields near the black hole's event horizon. They are primarily observed emanating from supermassive black holes at the centers of galaxies, where the intense gravitational pull accelerates matter to near-light speeds, creating the plasma jet. These ion jets are bright, elongated beams of plasma twisting into complex structures due to magnetic field interactions. Some of the largest plasma jets discovered can stretch across millions of light-years, exceeding the size of their host galaxy. Studying plasma jets helps astronomers understand the dynamics around black holes, the processes of accretion disks, and the evolution of galaxies.

Plasma jet from a supermassive black hole

Plasma Filaments

Plasma filaments in space are thread-like structures made of charged gas that appear in many places, including our Sun, stars, nebulae, and galaxies. They are commonly found in the Sun as prominences, spicules, coronal streamers, and polar plumes. They are found in nebulae in interstellar dust clouds that contain plasma filaments. The aurora is made of plasma creating curtains of filaments that create the northern and southern lights. Plasma filaments are found in cometary tails, the interstellar medium, and form cosmic webs that connect stars and galaxies.

Plasma filaments are formed when a large-amplitude Alfvén wave is disturbed by thermal or density fluctuations. Strong radial magnetic fields can pinch plasma into filaments and generally follow Birkeland currents. Plasma filaments form the cosmic spiderweb-like structure of the super-galaxies and trace out the distribution of matter and dust while helping to power the motion of galaxies.

Filamentation (or filamentary structuring) is often seen in plasmas. It is created because plasma contains free electrons, making it highly electrically conductive — even more than metals. As charged particles readily move in a plasma, a ring of magnetic field forms around the current that can pinch it into filamentary current strands (pinched filaments). They may also be formed alongside double layers of plasmas (which further accelerate charge particles), and if the resulting electrons approach relativistic velocities (the speed of light) and if they encounter a magnetic field, they will spiral and emit synchrotron radiation that includes radio, optical (light), x-rays, and gamma rays. Filamentation may also cause chemical separation of elements through Marklund convection.

Plasma filaments are also called Birkeland currents, "field aligned currents", "plasma ropes", "plasma cables", "flux ropes", "current constrictions", and "plasma rays." Hannes Alfvén argued that the spontaneous formation of current filaments is a fundamental feature of magnetized plasmas in the laboratory as well as in space.

A "plasma magnetic rope," also called a "magnetic flux rope," is a structure in a plasma where magnetic field lines twist and wrap around a central axis, forming a helical pattern, essentially like a twisted rope, and containing plasma within it. These structures are crucial for understanding various phenomena in space physics, particularly solar eruptions and coronal mass ejections from the Sun. The magnetic field lines within a rope are tightly wound around a common axis, creating a cylindrical shape with a strong axial magnetic field component. These ropes can form in regions of strong plasma currents where magnetic field lines become twisted due to complex dynamics within the plasma.

In the Sun's corona, magnetic flux ropes are believed to be the primary driver behind large solar eruptions like CMEs, where a sudden release of stored magnetic energy propels a massive cloud of plasma into space. Spacecraft observations using magnetometers can detect the characteristic magnetic field signature of a flux rope by measuring the changing direction of the magnetic field as the spacecraft passes through it.

Magnetic Plasma Ropes Connect the Sun to Earth

A "plasma magnetic rope" connects the Sun to the Earth every eight minutes through magnetic transfer flux portal reconnection via a twisted, tube-like structure of magnetic field lines within the solar plasma, called a "flux rope," which can extend from the Sun's corona all the way to Earth's magnetosphere, essentially acting as a conduit for charged particles to travel between the two celestial bodies, particularly during solar eruptions like coronal mass ejections (CMEs). These ropes consist of spiraling magnetic field lines, similar to a corkscrew (or DNA), carrying a large electric current through their core. They form on the Sun's surface within the corona, often associated with sunspots and active regions, where strong magnetic fields are present. When these flux ropes become unstable, they can erupt, releasing a massive burst of energy and plasma into space, known as a CME.

If a CME is directed towards Earth, the magnetic flux rope can interact with our planet's magnetosphere, potentially causing disruptions to power grids and satellites, depending on the strength of the interaction.

Magnetic reconnection occurs when magnetic field lines from different sources "reconnect," releasing a sudden burst of energy and facilitating the transfer of plasma between regions. Solar wind is a constant stream of charged particles emitted by the Sun that carries the magnetic field out into space, allowing the flux ropes to travel towards Earth in a constant stream of plasma.

A "magnetic flux portal," also called a "flux transfer event," is a temporary connection that forms between the Earth and the Sun when the two planet's magnetic fields briefly merge, creating a pathway for charged particles to flow between them, essentially acting like a "portal" connecting the two celestial bodies. This happens roughly every eight minutes on the dayside of Earth, where the magnetic fields directly face each other. When Earth's magnetic field presses against the Sun's magnetic field, the lines of force can momentarily reconnect, creating a cylindrical portal-like structure that is roughly the diameter of Earth. Through this portal, high-energy particles from the Sun's solar wind can readily flow into Earth's magnetosphere recharging it. These portals are not permanent and constantly open and close in an instantaneous reconnection process. Essentially, you could say that the solar breath of the Sun, every few minutes, inflates the lung-like plasmasphere of the Earth. Plus, these magnetic ropes also "pull" Earth behind the Sun like a cable tow. Rudolf Steiner was the first person to point this out.

Auroral Electrojet

An electrojet is an electric current which travels around the Earth's ionosphere. There are three electrojets: one above the magnetic equator (the equatorial electrojet), and one each near the Northern and Southern Polar Circles (the auroral electrojets). Electrojets are Hall currents carried primarily by electrons at altitudes from 100 to 150 km. The term 'auroral electrojet' is the name given to the large horizontal currents that flow in the D and E regions of the auroral ionosphere. Although horizontal ionospheric currents can be expected to flow at any latitude where horizontal ionospheric electric fields are present, the auroral electrojet currents are remarkable for their strength and persistence.

The auroral electrojet plays a key role in channeling plasma from the plasmasphere into the Earth's atmosphere, which is the primary mechanism behind the formation of auroras (Northern Lights and Southern Lights). Essentially, the electrojet acts as a conduit for these charged particles to flow along the Earth's magnetic field lines and collide with atmospheric atoms arising from the Earth, creating the visible light display.

The plasma that creates auroras originates from the plasmasphere, a region surrounding Earth filled with charged particles trapped by the planet's magnetic field. The auroral electrojet is a strong electric current flowing in the ionosphere, particularly near the polar regions, which acts as a pathway for the plasma to move towards the Earth's atmosphere. Plasma follows the Earth's magnetic field lines, which funnel the charged particles towards the polar regions where they collide with atmospheric gases, producing colorful auroras with the ionized nitrogen, oxygen, and other elements.

There are two main factors in the production of the electrojet. First of all, the conductivity of the auroral ionosphere is generally larger than that at lower latitudes. Secondly, the horizontal electric field in the auroral ionosphere is also larger than that at lower latitudes. Since the strength of the current is directly proportional to the vector product of the conductivity and the horizontal electric field, the auroral electrojet currents are generally larger than those at lower latitudes. During magnetically quiet periods, the electrojet is generally confined to the auroral oval. However, during disturbed periods, the electrojet increases in strength and expands to both higher and lower latitudes. This expansion results from two factors, enhanced particle precipitation and enhanced ionospheric electric fields.

We have now heard in the presentations above what atomic theory and modern materialistic science has to say about many aspects of plasma and the areas where it is active. Of course, many of these gravity-based atomic scientists do produce data that we can examine with an open mind and come to our own conclusions. Suffice it to say, plasma is much sought after and little understood by scientism, and should become a major area of study in life sciences to discover the connections between plasmas on Earth, in the human body, and in the cosmos.

To help us visualize the Earth's atmospheres in relationship to plasma formations around the Earth, we offer the chart below for quick reference.

Earth's Atmospheric Layers

Troposphere

- Earth's troposphere extends from Earth's surface to, on average, about 12 kilometers in height, with its height lower at Earth's poles and higher at the equator. Yet this very shallow layer is tasked with holding all the air that plants need for photosynthesis and animals need to breathe, and also contains about 99% of all water vapor and aerosols (minute solid or liquid particles suspended in the atmosphere).

- The temperature in the troposphere also decreases with height. On top of this layer is what is referred to as the tropopause.
- It is the densest atmospheric layer, compressed by the weight of the rest of the atmosphere above it.
- Most of Earth's weather happens here, and almost all clouds that are generated by weather are found here. Most aviation takes place here, including in the transition region between the troposphere and the stratosphere.

Stratosphere

- Located between approximately 12 and 50 kilometers above Earth's surface, the stratosphere is perhaps best known as home to Earth's ozone layer.
- In this region, the temperature increases with height. Heat is produced in the process of the formation of Ozone and this heat is responsible for temperature increase.
- It is nearly cloud- and weather-free, but polar stratospheric clouds occur mainly at high latitudes during the winter and are sometimes present in its lowest, coldest altitudes. It's also the highest part of the atmosphere that jet planes can reach.

Mesosphere

- Located between about 50 and 80 kilometers above Earth's surface, the mesosphere gets progressively colder with altitude. The top of this layer is the coldest place found within the Earth system, with an average temperature of about minus 85 °C (-120 °F).
- The very scarce water vapor present at the top of the mesosphere forms noctilucent clouds, the highest clouds in Earth's atmosphere.
- Most meteors burn up in this atmospheric layer. Sounding rockets and rocket-powered aircraft can reach the mesosphere. The transition boundary which separates the mesosphere from the stratosphere is called the stratopause.

Thermosphere

- It is located between about 80 and 700 kilometers above Earth's surface, whose lowest part contains the ionosphere. Because this layer is much closer to the sun, it can reach temperatures up to 2,000 °C (3,600 + °F)
- In this layer, temperatures increase with altitude due to the very low density of molecules found here. It is both cloud- and water-vapor-free.
- The aurora borealis (Northern lights) and aurora australis (Southern lights) are sometimes seen here. The International Space Station (ISS) orbits in the thermosphere.

Ionosphere

- It is not a distinct layer as the others mentioned above. Instead, the ionosphere overlaps the mesosphere, thermosphere, and exosphere.
- It is a very active part of the atmosphere and it grows and shrinks depending on the energy it absorbs from the Sun.
- It is an electrically conducting region capable of reflecting radio signals back to Earth.
- The electrically charged atoms and molecules that are formed in this way are called ions, giving the ionosphere its name and endowing this region with some special properties.

Exosphere

- It is located between about 700 and 10,000 kilometers above Earth's surface, the exosphere is the highest layer of Earth's atmosphere and, at its top, merges with the solar wind.
- Molecules found here are of extremely low density, so this layer does not behave like a gas, and particles here escape into space.
- While there's no weather in the exosphere, the aurora borealis and aurora australis are sometimes seen in its lowest part. Most Earth satellites orbit in this layer.

Are UFOs Actually Plasmoids?

"And it came to pass, as they still went on, and talked, that, behold, there appeared a chariot of fire, and horses of fire, and parted them both asunder; and \Elijah went up by a whirlwind into heaven."

2 Kings 2:11

Research Demonstrating the UFO Nature of Plasmoids

Extraterrestrial Life in the Thermosphere: Plasmas, UAP, Pre-Life, Fourth State of Matter

From: *Journal of Modern Physics*, Volume 15, Number 3, February 2024, R. Joseph, C. Impey, O. Planchon, R. del Gaudio, M. Abu Safa, A. R. Sumanarathna, E. Ansbro, D. Duvall, G. Bianciardi, C. H. Gibson, R. Schild

https://www.youtube.com/watch?v=zqMPh0fyuNk

Abstract

Plasmas up to a kilometer in size and behaving similarly to multicellular organisms have been filmed on ten separate NASA space shuttle missions, over 200 miles above Earth within the thermosphere. These self-illuminated "plasmas" are attracted to and may "feed on" electromagnetic radiation. They have different morphologies: 1) cone, 2) cloud, 3) donut, 4) spherical-cylindrical; and have been filmed flying towards and descending from the thermosphere into thunderstorms; congregating by the hundreds and interacting with satellites generating electromagnetic activity and approaching the Space Shuttles. Computerized analysis of flight path trajectories documents these plasmas travel at different velocities from different directions and change their angle of trajectory making 45°, 90°, and 180° shifts and follow each other. They have been filmed accelerating, slowing down, stopping, congregating, engaging in "hunter-predatory" behavior and intersecting plasmas leaving a plasma dust trail in their wake. Similar life-like behaviors have been demonstrated by plasmas created experimentally. Plasmas may have been photographed in the 1940s by WWII pilots (identified as "Foo fighters"); repeatedly observed and filmed by astronauts and military pilots and classified as Unidentified Aerial Anomalous Phenomenon. Plasmas are not biological but may represent a form of pre-life that via the incorporation of elements common in space, could result in the synthesis of RNA. Plasmas constitute a fourth state of matter, are attracted to electromagnetic activity, and when observed in the lower atmosphere likely account for many of the UFO-UAP sightings over the centuries.

1. Introduction: UAP and Extraterrestrial Electromagnetic Plasmas

Interstellar space and the upper atmosphere of Earth are permeated by plasmas. Plasmas are believed to constitute a fourth state of matter and may represent a form of pre-life or inorganic non-biological life. As detailed in this report, pulsating glowing "plasmas" up to a kilometer in size, and behaving similarly to simple multicellular organisms, have been filmed over 200 miles above Earth within the thermosphere, by ten separate space shuttle missions STS-48, STS-75, STS-80, STS-96, STS-101, STS-106, STS-115, STS-119, STS-123. These plasmas descend into the lower atmosphere of Earth, and when observed may have been classified as an "Unidentified Aerial Phenomenon" (or as some prefer "Anomalous Phenomenon").

As documented in this report, these plasma-like entities have been photographed congregating in the thermosphere above and diving into massive thunderstorms below; gathering around satellites generating electromagnetic pulses into the space medium; approaching and appearing outside the windows of NASA's Space Shuttles; changing velocity and direction and making sudden 45 degree to 90 degree turns, and following and even colliding and intersecting one another.

These glowing pulsating plasmas in the thermosphere grow in size, replicate, are attracted to, and make contact and intersect one another often releasing kilometers-in-length plasma-dust trails in their wake. Yet others engage in what appears to be hunter-predatory behavior; tracking, then accelerating and piercing other plasmas that may be several dozen kilometers distant, even striking and intersecting up to ten or more at varying distances from one another. They display all the characteristics of life-like plasmas as defined by Alfvén, Teordorani, Lozneanu and Sanduloviciu and Tsytovich, et al.

Although plasmas in the thermosphere engage in behavior, in all respects, plasmas are not biological entities, and their actions are mediated by electromagnetic and other non-biological factors including the incorporation of radiated dust. Plasmas may have high or low density, high or low temperatures, may be stable or unstable, and consist of positively and negatively charged particles, ionized atoms of gas; and whose basic interactions are electromagnetic.

In the upper atmosphere, plasmas are believed to react and possibly form in response to intermittent turbulence, geomagnetic storms, coronal mass ejections, solar flares, eclipses, the waxing and waning of sunlight, atmospheric waves, radiation from radio transmitters and heating facilities, and the incorporation of dust; all which affect the shape, velocity, and behavior of plasmas.

As detailed by Teodorani Ivan et al. Lozneanu, Sanduloviciu Tsytovich, et al, plasmas can form complex life-like shapes as well as a nucleus that may acquire DNA-like capabilities. For example, space is permeated by dust; and when a plasma incorporates dust grains, the plasma acquires an electric charge that sucks in electrons which attract positively charged ions giving rise to "plasma crystals" which also contain organic matter including fragments of carbonaceous chondrites.

Electrostatic forces and the polarization of the plasma cause these plasma dust crystals to twist, spin, and form helical structures that can evolve into a double helix similar to the double helix of DNA. Teodorani argues that these helixes can then be replicated in other crystals and that the self-organizing ability of these crystals can affect other plasmas which may also form a plasma dust-crystal helix.

These extraterrestrial "plasmas" filmed in the thermosphere may represent a non-biological form of pre-life before the acquisition of RNA and then DNA. Further, when they descend into the lower atmosphere they would likely be classified as UFOs and UAP when observed.

Non-equilibrium Formation and Relaxation of Magnetic Flux Ropes at Kinetic Scales

Communications Physics, Volume 7, Article Number: 297, 2024, by Young Dae Yoon, Modhuchandra Laishram, Thomas Earle Moore & Gunsu S. Yun

Abstract

Magnetic flux ropes are pivotal structures and building blocks in astrophysical and laboratory plasmas, and various equilibrium models have thus been studied in the past. However, flux ropes in general form at non-equilibrium, and their pathway from formation to relaxation is a crucial process that determines their eventual properties. Here we show that any current parallel to a background magnetic field will evolve into a flux rope via non-equilibrium processes. The detailed kinetic dynamics are exhaustively explained through single-particle and Vlasov analyses and verified through

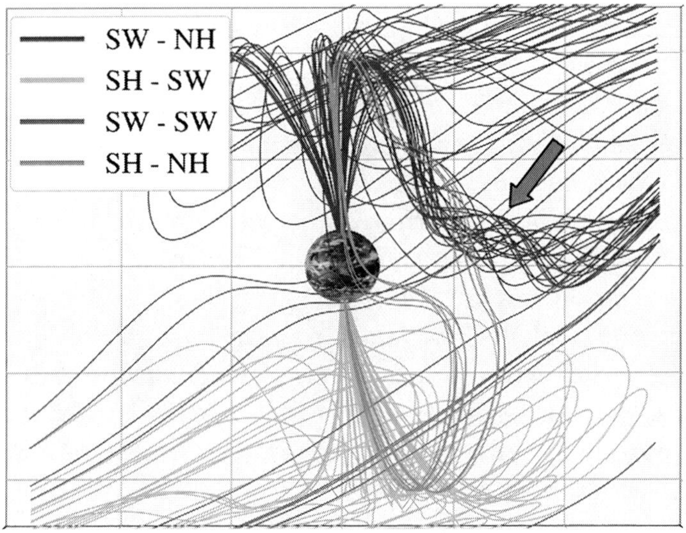

Magnetic field lines

particle-in-cell simulations. This process is consistent with many proposed mechanisms of flux rope generation such as magnetic reconnection.

Localized magnetic reconnection at the dayside magnetopause leads to the production of Flux Transfer Events (FTEs). The magnetic fields within the FTEs exhibit complex helical flux-rope topologies.

Magnetic flux ropes are twisted bundles of magnetic field lines that confine current-carrying plasma and are fundamental and ubiquitous structures in space, astrophysical, and laboratory plasmas. They act as underlying structures in various plasma phenomena and instabilities, serving as magnetic batteries that convert magnetic energy into other forms of energy via processes such as magnetic reconnection, eruption, flux transfer events (FTEs), and current-driven/kinetic instabilities. Due to their significance, various equilibrium models such as force-free models, magnetohydrodynamic (MHD) models, and MHD models including deformation effects have been developed to explain various observed characteristics of these structures.

Despite extensive studies, no single model can comprehensively interpret flux ropes. For example, only 60% of flux ropes observed in the magnetotail can be described by force-free models. Thus, there is a need to revisit the problem of whether all flux ropes are alike in terms of morphology, magnetic, and plasma properties and dynamics. Answering this question calls for a revelation of their formation, relaxation, and evolution. In particular, for sub-ion-scale systems, the detailed structure of the particle distribution function is an important determining factor of their stability and subsequent dynamics.

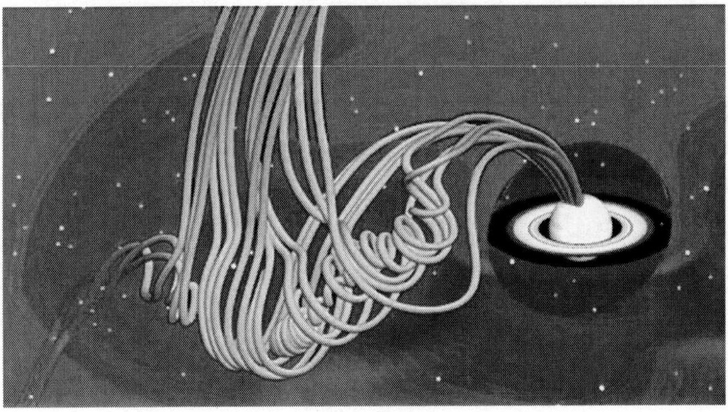

Birkeland magnetic currents

Plasma Spheromaks

A spheromak is an arrangement of plasma formed into a toroidal shape similar to a smoke ring. The spheromak contains large internal electric currents and their associated magnetic fields arranged so the magnetohydrodynamic forces within the spheromak are nearly balanced, resulting in long-lived (microsecond) confinement times without external fields. Spheromaks belong to a type of plasma configuration referred to as compact toroids. A spheromak can be made and sustained using magnetic flux injection, leading to a dynomak.

The physics of the spheromak and of collisions between spheromaks is similar to a variety of astrophysical events, like coronal loops and filaments, relativistic jets, and plasmoids. They are particularly useful for studying magnetic reconnection events, when two or more spheromaks collide. Spheromaks are easy to generate using a "gun" that ejects spheromaks off the end of an electrode into a holding area, called the flux conserver. This has made them useful in the laboratory setting, and spheromak guns are relatively common in astrophysics labs.

Spheromaks have been proposed as a magnetic fusion energy concept due to their long confinement times, which was on the same order as the best tokamaks when they were first studied. Although they had some successes during the 1970s and 1980s, these small and lower-energy devices had limited performance and most spheromak research ended when fusion funding was dramatically curtailed in the late 1980s. However, in the late 1990s research demonstrated that hotter spheromaks have better confinement times, and this led to a second wave of spheromak machines. Spheromaks have also been used to inject plasma into a bigger magnetic confinement experiment like a tokamak.

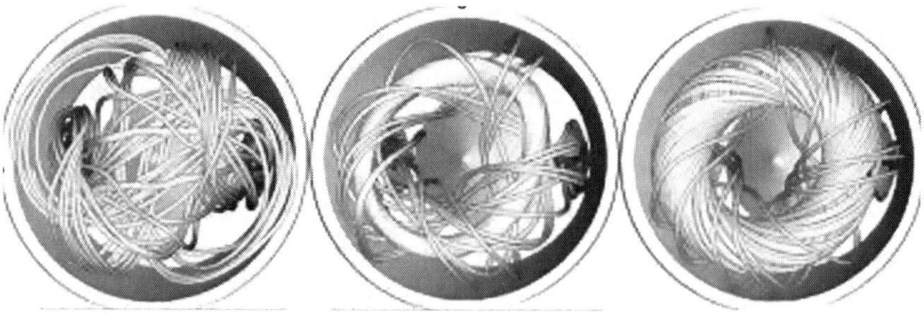

A model of a spheromak forming inside a chamber (far right)

Marklund Convection

The filaments of light streaming from the Sun self-organize individual elements along axially aligned sheaths. This is a well-known observation in cosmology and is known as markland convection. Charge collapses into concentric, collimated, cylindrical, counter-rotating tubes within tubes. Matter self-organizes in these filaments of markland convection.

Marklund convection was named after Swedish physicist Göran Marklund and is a convection process that takes place in filamentary currents of plasma. It occurs within a plasma with an associated electric field that causes convection of ions and electrons inward towards a central twisting filamentary axis. A temperature gradient within the plasma will also cause chemical separation based on different ionization potentials.

Plasma convects radially inwards towards the center of a cylindrical flux tube. During this convection, the different chemical constituents of the plasma, each having its specific ionization potential, enters into a progressively cooler region. The plasma constituents will recombine and become neutral, and thus no longer under the influence of the electromagnetic forcing. The ionization potentials will thus determine where the different chemicals will be deposited. This provides an efficient means to accumulate matter within a plasma. In a partially ionized plasma, electromagnetic forces act on the non-ionized material indirectly through the viscosity between the ionized and non-ionized material.

Hannes Alfvén showed that elements with the lowest ionization potential are brought closest to the axis and form concentric hollow cylinders whose radii increase with ionization potential. The drift of ionized matter from the surroundings into the rope means that the rope acts as an ion pump, which evacuates surrounding regions, producing areas of extremely low density.

Plasma Portals to the Sun

A flux transfer event (FTE) occurs when a magnetic portal opens in the Earth's magnetosphere through which high-energy particles flow from the Sun. This connection, while previously thought to be permanent, has been found to be brief and very dynamic. The European Space Agency's four

Cluster spacecraft and NASA's five THEMIS probes have flown through and surrounded these FTEs, measuring their dimensions and identifying the particles that are transferred between the magnetic fields.

Earth's magnetosphere and the Sun's magnetic field are constantly pressed against one another on the dayside of Earth. Approximately every eight minutes, these fields briefly merge, forming a temporary "portal" between the Earth and the Sun through which high-energy particles such as solar wind can flow. The portal takes the shape of a magnetic cylinder about the width of Earth. Current observations place the portal at up to four times the size of Earth.

The Van Allen radiation belt is a zone of energetic charged particles, most of which originate from the solar wind, which are captured by and held around a planet by that planet's magnetosphere. Earth has two such belts, and sometimes others may be temporarily created. The belts are named after James Van Allen, who published an article describing the belts in 1958.

Earth's two main belts extend from an altitude of about 400 to 36,040 miles above the surface, in which region radiation levels vary. The belts are in the inner region of Earth's magnetic field. They trap energetic electrons and protons. Other nuclei, such as alpha particles, are less prevalent. Most of the particles that form the belts are thought to come from the solar wind while others arrive as cosmic rays. By trapping the solar wind, the magnetic field deflects those energetic particles and protects the atmosphere from destruction.

The inner Van Allen Belt extends typically from an altitude of 0.2 to 2 Earth radii or 620 miles to 7,500 miles above the Earth. In certain cases, when solar activity is stronger or in geographical areas such as the South Atlantic Anomaly, the inner boundary may decline to roughly 200

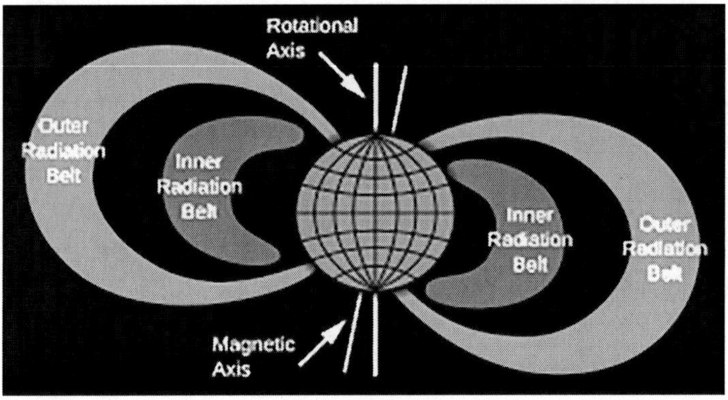

The Van Allen belts

km above the Earth's surface. The inner belt contains high concentrations of electrons in the range of hundreds of keV and energetic protons with energies exceeding 100 MeV—trapped by the relatively strong magnetic fields in the region (as compared to the outer belt).

The outer belt consists mainly of high-energy (0.1–10 MeV) electrons trapped by the Earth's magnetosphere. It is more variable than the inner belt, as it is more easily influenced by solar activity. It is almost toroidal in shape, beginning at an altitude of 3 Earth radii and extending to 10 Earth radii—8,100 to 37,300 miles above the Earth's surface. Its greatest intensity is usually around 4 to 5 RE. The outer electron radiation belt is mostly produced by inward radial diffusion and local acceleration due to transfer of energy from whistler-mode plasma waves to radiation belt electrons. Radiation belt electrons are also constantly removed by collisions with Earth's atmosphere, losses to the magnetopause, and their outward radial diffusion. The gyro-radii of energetic protons would be large enough to bring them into contact with the Earth's atmosphere. Within this belt, the electrons have a high flux and at the outer edge (close to the magnetopause), where geomagnetic field lines open into the geomagnetic "tail", the flux of energetic electrons can drop to the low interplanetary levels within about 62 miles—a decrease by a factor of 1,000.

Plasma Weapons

"He drove out the man, and at the east of the garden of Eden he placed the cherubim and a flaming sword that turned every way to guard the way to the Tree of Life."

Genesis 1-3:24

Do Plasma Guns exist? Yes, if you define a plasma gun as an object that shoots out plasma, then yes, plasma guns have existed for quite a while. They come in many forms including a plasma torch, gun, laser, or Tesla coil. Plasma guns are one of several United States Government efforts to develop plasma-based projectiles. The first computer simulations occurred in 1990, and its first published experiment appeared on 1 August 1993. A plasma cannon (also called an electrothermal accelerator) is an projectile weapon, which accelerates a plasma projectile by means of a plasma discharge between electrodes at the rear of the barrel, generating a rapid increase in pressure.

This weapon is the 'tip of the iceberg' when considering weaponized plasma. The long lists below are only a few weapon systems that have been made public by the warmongers who spend billions on these death toys. What the Department of Defense currently has as an assortment of plasma weapons is top secret. But you can see the direction these unregulated innovators are going with the "fabric of the universe" that they believe is a weapon to destroy the universe. All plasma weapons should be banned and a moratorium called on plasma weapons until proper research and regulation finds them safe.

Shiva Star

https://www.bbc.com/news/technology-18630622

Shiva Star at the Air Force Research Laboratory at Kirtland Air Force Base. Shiva Star is a high-powered pulsed-power research device located at the Air Force Research Laboratory on the Kirtland Air Force Base in Albuquerque, New Mexico. The device was originally built in the 1970s for high-power x-ray research, was later re-directed to study the Strategic Defense Initiative (SDI) and is now being used for magnetized target fusion research. Shiva Star was also used to develop an experimental weapon known as MARAUDER for the SDI effort between 1989 and 1995. The idea appears to have been to create compact toroids of high-density plasma that would be ejected from the device using a massive magnetic pulse. The plasma projectiles would be shot at a speed expected to be 3000 km/s in 1995 and 10,000 km/s (3.3% of the speed of light) by 2000. A shot has the energy of 2.27 kg of TNT exploding. Doughnut-shaped rings of plasma and 'balls of lightning' exploded with devastating thermal and mechanical effects when hitting their target and produced a pulse of electromagnetic radiation that could scramble electronics. The project became classified, and no information about the fate of the project has been published after 1995.

MARAUDER

MARAUDER (Magnetically Accelerated Ring to Achieve Ultra-high Directed Energy and Radiation) was a United States Air Force Research Laboratory project concerning the development of a coaxial plasma railgun. The first computer simulations occurred in 1990, and its first published experiment appeared on August 1, 1993. The first MARAUDER experiment was motivated by RACE, (the Ring Accelerator Experiment) which took place in 1991 at Lawrence Livermore National Laboratory. The ultimate goal of the MARAUDER program was to accelerate highly dense toroids containing plasma to high speeds. Such a system could be used for "hyper-velocity projectiles," x-ray production, and electrical power amplification. The stated goals of the program included studying the "formation, compression, and acceleration of magnetized plasma rings." Specifically, the objective of the program was the acceleration of a toroid of 0.5 to 2 milligrams (0.0077 to 0.0309 gr) of plasma into a kinetic energy level on the order of megajoules using a 5-10 MJ coaxial gun design.

In 1993, the Phillips Laboratory developed a coaxial plasma gun capable of accelerating 1 to 2 milligrams (0.015 to 0.031 gr) of plasma by up to 100 billion g (9.8×1011 m/s2) in a toroid of 1 meter in diameter. The toroids are similar to spheromaks but differ in that an inner conductor is used to accelerate the plasma and that confinement behavior results from interactions of the toroid with its surrounding atmosphere. It utilized the Shiva Star capacitor bank to satisfy the large energy requirements that were needed.

The plasma projectiles would be shot at a speed expected to be 3,000 kilometers per second (1,900 mi/s) in 1995 and 10,000 kilometres per second (6,200 mi/s) (3% of the speed of light) by 2000. A shot has the energy of 5 pounds of TNT (2.27 kilograms of TNT (9.5 MJ)).

Hammer, Hartman et al. showed that spheromaks could be accelerated to extremely high velocities using a railgun, which led to several proposed uses. Among these was the use of such plasmas as "bullets" to fire at incoming warheads with the hope that the associated electric currents would disrupt their electronics.

Lightning Laser Weapon Developed by US Army, Published June 28, 2012

https://www.amazon.com/Artificial-Lightning-Generator-Experiment-Education/dp/B0CQG369W5

Laser-guided lightning gun is a weapon capable of emitting huge power which can fire a laser-guided lightning bolt at a target. The Laser-Induced Plasma Channel (LIPC) is designed to hit targets that conduct electricity better than the air or ground that surrounds them. Details of the weapon were released on the US Army's website. If a laser beam is intense enough, its electro-magnetic field is strong enough to rip electrons off of air molecules, creating plasma. This plasma is located along the path of the laser beam to direct it by moving a mirror.

Tesla Coil Handheld Gun, Artificial Lightning, Portable Mini Spark Gap Arc Generator

- The arc generated by this equipment has a certain high temperature, which can ignite some flammable and explosive materials, do not use this equipment in places where flammable and explosive materials are placed or in storage warehouses.

- When the device is working, it will generate certain magnetic and electric field interference, therefore, people who have implanted or wear precision medical devices in the body should not use or be close to the device that is working.

Plasma Ghosts

https://hackaday.com/2020/05/18/plasma-ghosts-may-help-keep-future-aircraft-safe/

In an effort to develop the ultimate in defensive countermeasures, the United States Navy has been working on a system that can project decoy aircraft in mid-air. Referred to as "Ghosts" in the recently published patent, several of these phantom aircraft could be generated for as long as the system has electrical power.

It's probably best to think of the projection as a solid-state replacement to the traditional flare: the idea is to create a mass of plasma in the air that is emitting enough energy at the appropriate wavelengths that it would distract from the physical aircraft it's being projected from. But that doesn't necessarily mean the projection will be an amorphous blob, either. The patent specifically mentions that the system's

248 nm krypton fluoride (KrF) laser could be paired with mirrors and raster optics to create 2D and 3D volumetric images in space. It goes on to compare the technique with the way images were drawn with a scanning electron beam in a cathode-ray tube television.

Another key improvement is that the plasma decoy will actually follow the aircraft as it flies. Corporeal countermeasures are quickly left behind, but these ghosts would be able to maintain a set distance from the aircraft throughout the entire mission.

As it so happens, the technology to project relatively simple geometric shapes in the air with a laser isn't exactly new. A Japanese company demonstrated using laser-induced plasma balls as "pixels" in a rudimentary display as early as 2011.

The concept is also very similar to research being conducted by the Department of Defense into open-air plasma speakers. Developed as part of the Joint Non-Lethal Weapons Program, the light and sound produced by the plasma ignition was envisioned as a way to stun or confuse a target. One demonstration even showed how the plasma ball could be modulated in such a way that they could emit an audible warning message.

U.S. Navy Laser Creates Plasma 'UFOs', Updated April 14, 2022

https://www.forbes.com/sites/davidhambling/2020/05/11/us-navy-laser-creates-plasma-ufos/

The U.S. Navy has patented technology to create mid-air images to fool infrared and other sensors. This builds on many years of laser-plasma research and offers a game-changing method of protecting aircraft from heat-seeking missiles. A sufficiently intense laser pulse can ionize producing a burst of glowing plasma. The Laser Induced Plasma Effects program uses single plasma bursts as flash-bang stun grenades; a rapid series of such pluses can even be modulated to transmit a spoken message. In 2011, Japanese company Burton Inc. demonstrated a rudimentary system that created moving 3D images in mid-air with a series of rapidly generated plasma dots.

A more sophisticated approach uses an intense, ultra-short, self-focusing laser pulse to create a glowing filament or channel of plasma, an effect discovered in the 1990s. Known as laser-induced plasma filaments (LIPF) these can be created at some distance from the laser for tens or hundreds of meters. Because LIPFs conduct electricity, they have been investigated as a means of triggering lightning or creating a lightning gun.

The patent goes on to explain that the laser creates a series of mid-air plasma columns, which form a 2D or 3D image by a process of raster scanning, similar to the way old-style cathode ray TVs sets display a picture. Unlike flares, the LIPF decoy can be created instantly at any desired distance from the aircraft and can be moved around at will.

If you have a very short pulse, you can generate a filament, and in the air that can propagate for hundreds of meters, and maybe with the next generation of lasers you could produce a filament miles long.

Other nations may have their own version. In the early 1990s the Russians claimed that they could produce glowing 'plasmoids' at high altitude using high-power microwave or laser beams; these were intended to disrupt the flight of ballistic missiles, an answer to the planned American 'Star Wars.'

Plasma Guns

https://aamsworld.com/nos-solutions/plasma/?lang=en

The Plasma Gun® is the first polarizing plasma generator which has a large application surface. With three diffusion modes: hot, cold and hybrid, the Plasma Gun® masters all the properties of plasma to clean up, restructure, improve or repair the skin in a non-surgical way. Particularly effective on the epidermis, Plasma Gun® has an anti-ageing effect on wrinkles, stretch marks, scar marks and light to moderate skin imperfections.

Patented technology, the Plasma Gun® uses a tip that generates a micro-arc by ionization of the ambient air. The plasma stimulates the production of collagen, elastin, and fibroblasts. Used on most parts of plasma devices, traditional plasma consists in emitting hot plasma continuously without electronic management of the dose. The user must manage by himself the exposure time on the skin and therefore the intensity of the emission.

The Plasma Gun® also generates cold plasma through a patented DBD C® (Dielectric Barrier Discharge Cold®) tip especially created for this application. It combines specific materials studied components to conserve all the temperature resulting in order to diffuse all the Plasma energy on a wide area. The Plasma Gun® emits cold plasma in order to distribute all the plasma energy over a large surface area while maintaining the temperature generated by this energy without emitting a sensation of direct heat on the skin.

Dense Plasma Focus

A dense plasma focus (DPF) is a type of plasma generating system originally developed as a fusion power device starting in the early 1960s. The original concept was developed in 1954 by N.V. Filippov, who noticed the effect while working on early "pinch machines" in the USSR. A major research program on DPF was carried out in the USSR through the late 1950s and continues to this day. A different version of the same basic concept was independently discovered in the US by J.W. Mather in the early 1960s. This version saw some development in the 1970s, and variations continue to be developed.

The basic design derives from the z-pinch concept. Both the DPF and pinch use large electrical currents run through a gas to cause it to ionize into plasma and then pinch down on itself to increase the density and temperature of the plasma. The DPF differs largely in form; most devices use two concentric cylinders and form the pinch at the end of the central cylinder. In contrast, z-pinch systems generally use a single cylinder, sometimes a torus, and pinch the plasma into the center.

The plasma focus is similar to the high-intensity plasma gun device (HIPGD) (or just plasma gun), which ejects plasma in the form of a plasmoid, without pinching it. A comprehensive review of the Dense Plasma Focus and its diverse applications has been made by Krishnan in 2012.

Magnetohydrodynamic Generators

MHD generators are different from traditional electric generators because they can operate at high temperatures without moving parts. They work by converting the kinetic energy of fast-moving ionized gases into electricity using a strong magnetic field. The principle of MHD power generation is based on Faraday's law of electromagnetic induction.

The MHD disk generator has been used to study plasma properties and develop new diagnostic techniques. It has been integrated into electrical power plants, and studies show it can achieve efficiencies comparable to linear generators.

Liquid Metal MHD (LMMHD) power systems are being proposed for direct electrical energy conversion of low-grade thermal sources, such as solar energy. They are attractive for their efficiency and cost per unit of installed power.

Seeding alternatives are ongoing to identify alternatives to the use of seed materials to improve electrical field strength. The goal is to identify non-seeded approaches that improve overall power plant design configurations. Fluid dynamics and plasma interfaces include developing more accurate and faster tools for modeling systems and validating concepts experimentally.

Federally Sponsored Research and Development

https://patents.google.com/patent/US20200041236A1/en
Application US16/055,684 events—2018-08-06
Application filed by US Department of Navy—2018-08-06
Priority to US16/055,684—2018-08-06
Assigned to Unite States of America as Represented by the Secretary of the Navy

Laser induced plasma emission spectra covers a wide electromagnetic spectrum, from Infrared (IR) to Visible (VIS) and up to Ultraviolet region (UV). By fine-tuning the interaction parameters (e.g. laser wavelength, laser temporal and spatial pulse profile, etc.) it is possible to maximize the radiating power for a dedicated electromagnetic spectrum.

Once the presence of an activated IR missile is indicated, flares are released in an attempt to decoy the missile; some systems are automatic, while others require manual jettisoning of the flares. Flares burn at thousands of degrees, which is much hotter than the exhaust of a jet engine. IR missiles seek out the hotter flame, believing it to be an aircraft in afterburner or the beginning of the engine's exhaust source.

Described herein is a system and method to generate a plasma-based decoy flare by using a laser source, to counter an infrared homing surface-to-air and/or air-to-air missile. With laser-induced plasma (LIP), it is possible to generate multiple wavelengths just by "tuning" the laser parameters. This method allows for an ultra-fast response time. Due to the fact that the effect is generated by the laser beam interaction with air, the time required to produce the flares is less than a millionth of a second.

Through optical beam forming techniques, an array of plasma columns can be created, forming a sheet-like plasma, creating a layer of excited electrons in the air. This layer can be used as a reflective surface, or mirror, for incident energies whose frequencies are below the plasma frequency, reflecting the power away from the intended path. The layer can also be used instead to deflect, diffract, or redirect the incident energy in a different direction.

By rastering plasma, it is possible to generate a 2D or 3D volumetric image in space. This is analogous to the rastering of an electron beam in a cathode ray tube-based television. In one potential embodiment, a laser system would be mounted on the back of an air vehicle such that the beam can be rastered using optics and mirrors to generate a large 'ghost' image in space. This 'ghost' image would appear to detract the homing missile away from the tangible air vehicle.

LIP can be generated using a 248 nm KrF excimer laser. In addition to a LIP, any other type of laser or light and/or electromagnetic source can be used as a decoy in this manner, including radio frequency (RF) and Microwave generators, High Power Lasers (HEL) and High Power LEDs. Depending on the desired use, the pulse characteristics of the electromagnetic source (energy, pulse shape, duration, repetition rate) are critical in achieving the required plasma parameters. A laser source is mounted on the back of an air vehicle with mirrors and optics that enable the raster scanning of laser source to create LIP, which acts as a virtual 'ghost' object. Alternatively, LIP could also be manipulated and distributed using a laser gimbal or turret which can be easily installed on air vehicles, which could be anything from an (aircraft, helicopter, ship, etc.).

An LIP flare array propagates in air at the speed of light, allowing for immediate deployment of a countermeasure to protect against an incoming threat. The potential applications of this LIP flare/decoy can be expanded, such as using a helicopter deploying flares to protect a battleship or using this method to cover and protect a whole battle-group of ships, a military base or an entire city.

Similar Patents for Weaponized Plasma

- US11359892B2—2022-06-14—System and method for laser-induced plasma for infrared homing missile countermeasure
- US7930967B2 20—11-04-26—Method for antimissile protection of vehicles and implementing device
- US8339580B2—2012-12-25—Sensor-guided threat countermeasure system
- US20160097616A1—2016-04-07—Laser Guided and Laser Powered Energy Discharge Device
- US10508889B1—2019-12-17—Method and apparatus for laser-induced plasma filaments for agile counter-directed energy weapon applications
- US7943914B2 —2011-05-17—Back illumination method for counter measuring IR guided missiles

- AU2016245376A1—2017-10-26—A weapons counter measure method and apparatus
- Steinvall—2021—The potential role of laser in combating UAVs: Part 2; laser as a countermeasure and weapon
- Forden—1997—The airborne laser
- Weise et al.—2004— Overview of directed energy weapon developments
- US5401976A—1995-03-28—Process to camouflage heat emitting device and particle for process
- Olson—2012—History of laser weapon research
- CN112888972A—2021-06-01—System and method for laser scattering, deflection, and manipulation
- US20230087750A1—2023-03-23—Plasma Burst Application System and Method
- KR101962271B1—2019-03-26—Infrared plume simulator
- US12104885B2—2024-10-01—Laser induced plasma targets for use as an electromagnetic testbed
- RU2790364C1—2023-02-17—Method for Laser Destruction of UAVs by the System
- US12140726B2—2024-11-12—Remote monostatic tomography system and related methods
- Solin—2014— Airborne laser threat to commercial space telescopes

- RU214292U1—2022-10-19 - Means of individual protection of an aircraft against guided missiles with an optical homing head
- US20240230946A1—2024-07-11—Remote monostatic tomography system and related methods
- Patel et al.—1987—Strategic defense and directed-energy weapons
- US20240248295A1—2024-07-25—Telescope with stiffening rib
- Carter et al.—2022—DE Weapons, MASERS/LASERS
- Braid et al.—1990—Laser brightness verification
- US8827211B2—2007-10-23—2014-09-09—Laser-based flow modification to remotely control air vehicle flight path
- US20160097616A1—2011-11-25—2016-04-07—Laser Guided and Laser Powered Energy Discharge Device
- US8981261B1—2012-05-30—2015-03-17—The Boeing Company Method and system for shockwave attenuation via electromagnetic arc
- US11359892B2—2018-08-06—2022-06-14—United States of America as Represented by the Secretary of the Navy, System and method for laser-induced plasma for infrared homing missile countermeasure
- US11131685B2—2019-05-15—2021-09-28—The Boeing Company—Air data system using magnetically induced voltage

Research with Plasma Generators

Bring me my Bow of burning gold: Bring me my arrows of desire:
Bring me my Spear: O clouds unfold! Bring me my Chariot of Fire!

William Blake, *Jerusalem*

MHD Generation for Sustainable Development, from Thermal to Wave Energy Conversion

José Carlos Domínguez-Lozoya, ORCID, David Roberto Domínguez-Lozoya, Sergio Cuevas, and Raúl Alejandro Ávalos-Zúñiga

Abstract

Magnetohydrodynamic (MHD) generators are direct energy conversion devices that transform the motion of an electrically conducting fluid into electricity through interaction with a magnetic field. Developed as an alternative to conventional turbine-generator systems, MHD generators evolved through the 20th century from large units, which are intended to transform thermal energy into electricity using plasma as a working fluid, to smaller units that can harness heat from a variety of sources. In the last few decades, an effort has been made to develop energy conversion systems that incorporate MHD generators to harvest renewable sources such as solar and ocean energy, strengthening the sustainability of this technology. This review briefly synthesizes the main steps in the evolution of MHD technology for electricity generation, starting by outlining its physical principles and the proposals to convert thermal energy into electricity, either using a high-temperature plasma as a working fluid or a liquid metal in a one- or two-phase flow at lower temperatures. The use of wave energy in the form of acoustic waves, which were obtained from the conversion of thermal energy through thermoacoustic devices coupled to liquid metal and plasma MHD generators, as well as alternatives for the transformation of environmental energy resources employing MHD transducers, is also assessed. Finally, proposals for the conversion of ocean energy, mainly in the form of waves and tides, into electric energy, through MHD generators using either seawater or liquid metal as working fluids, are presented along with some of the challenges of MHD conversion technology.

Plasma Generators in the Human Body, The Gospel of Sophia, Volume 2, Dr. Tyla Gabriel

In the previous chapter, Tibetan and Vajrayogini practices showed that an aspirant could open his thousand-petaled crown chakra and begin the path of ascension by following certain exercises or tantric yoga. Although they may seem like mythical, fantastical images, we can find similarities to modern technologies of energy production called magnetohydrodynamic generators. Once understood, these generators can provide the aspirant with a functional example of how the earthly and cosmic nutrition streams and the etherization of the blood occurs in the human body and, consequently, in one's own spiritual development. In the same way that we have drawn connections between different mystery traditions in this book, we also are mindful to connect the dots with any practices, processes, or tools—both ancient and modern—that can help illuminate the path for the aspirant's spiritual development.

Plasma generators are some of the most powerful direct current generators in use today. To produce direct current, the generators stream a heavy metal-impregnated gas/plasma (a conductor) across an open magnetic gap. In a closed system, they will keep going once started and continue as long as the magnetic field holds and the gas/plasma keeps passing over the gap in the magnetic field. Magneto-hydrodynamic generators are capable of transforming thermal energy and kinetic energy directly into electricity. Likewise, the magnetic energy of larger systems such as the earth and sun are generated in a similar fashion.

It may surprise many that the circulation of human blood throughout the body by the beating heart produces enough kinetic energy to establish a field of magnetic force in and around the heart and torso. As the blood stream passes through the heat of the lung/blood exchange, the presence of iron and other metals establishes the metallic fluid necessary to create this MHD dynamo. Blood also acts like a gas. What most of us cannot see—but clairvoyants can—is that near the heart is a flame about the size of a human thumb. A small furnace driven by the breathing process, the little flame arises from the interaction of the lung passing the warmth of oxygen to the heart. When the heat from the lungs creates this flame between the lung and heart, the metal-laden gas of the blood flows through this gap. Here, direct electrical current is established. Both electrostatic and electromagnetic forces create a torus of energy in a coherent field of bio-electromagnetic energy.

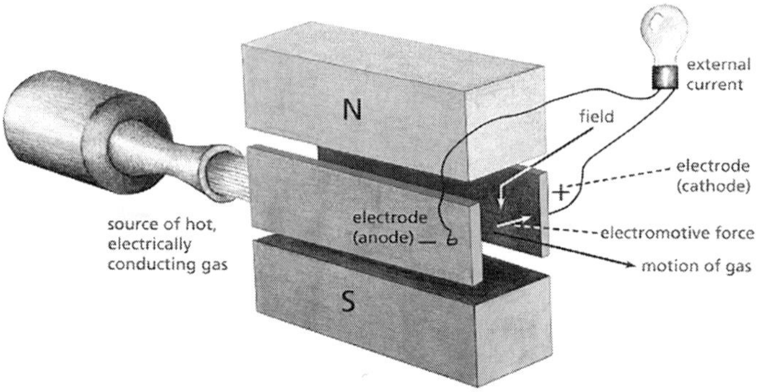

A simplified illustration of a magneto-hydrodynamic generator

The human heart is an energy dynamo with a plasma-generated magneto-hydrodynamic force. The energy created by this human MHD generator is far beyond the previous attempts to explain bioenergetics as a simple transfer of cellular energy. The human heart creates a field of magnetism that extends up to fifteen feet. This energy field is basically the center of three fields of force: one in the heart, one in the brain, and one that extends from the combined three lower chakras. These centers of energy production in the brain, heart, and lower abdomen relate to soul capacities that also find their centers of activity in these three areas. Thinking finds its home in the brain, feeling in the heart, and willing in the three lowest chakras.

It only takes modern science to examine the three centers of vital forces (thinking, feeling, and willing) to reveal a wealth of information about the resonant fields of morphic energy that are created by these plasma generators. Still little understood by science, the three centers generate a tremendous amount of energy. Although scientists may use instruments to determine accurate measurements of this energy, they still have not yet discovered its source and nature.

Interestingly, these fields of force have been found to resemble the Van Allen belts of the earth that create and maintain the magnetic sphere of the globe. These donut-shaped fields of force are called toroidal fields, which are the key mechanisms for thinking, feeling, and willing. When activated by consciousness through practices such as tantric yoga, meditation, prayer, or contemplation, these forces evolve to become the spiritual forces of Imagination, Inspiration, and Intuition—topics covered repeatedly throughout The Gospel of Sophia.

Here is the key: the processes whereby these toroidal fields of force come into being and are sustained are directly connected to the earthly and cosmic nutrition streams. They constitute the manifested energy that arises from the expansion of consciousness. When nourished properly, both physically and spiritually, these force fields are potentially unlimited, giving the aspirant the ability to attain unlimited spiritual power.

Manmade MHD Heart

An artificial heart using magnetohydrodynamic propulsion is an implantable artificial heart apparatus which has no moving parts. It employs the technology of magnetohydrodynamics (MHD) to induce flow of human blood. An MHD propulsion unit is a device which applies magnetic and electric fields to blood to propel the conductive fluid through the body's circulatory system. Electricity is passed through the blood via electrodes at the same junction where the blood is exposed to the magnetic field. Charged ions that move from anode to cathode create their own corresponding magnetic fields which are either attracted or repelled by the externally applied magnetic field. The result is propulsion in a uniform direction with the moving ions, in effect, dragging fluid molecules with them. A single MHD propulsion unit performs the combined functions of the atrium and the ventricle of the human heart. Two MHD propulsion units are required to perform the full cardiac cycle of the heart, one to pump unoxygenated blood through the lungs and one to pump oxygenated blood through the body's circulatory system.

From: The Heart Math Institute

The magnetic fields produced by the heart are involved in energetic communication, which we also refer to as cardio-electromagnetic communication. The heart is the most powerful source of electromagnetic energy in the human body, producing the largest rhythmic electromagnetic field of any of the body's organs. The heart's electrical field is many times greater in amplitude than the electrical activity generated by the brain. This field, measured in the form of an electro-cardiogram (ECG), can be detected anywhere on the surface of the body. Furthermore, the magnetic field produced by the heart is more than 100 times greater in strength than the field generated by the brain and can be detected up to three feet away from the body, in all directions, using SQUID-based magnetometers.

Prompted by our findings that the timing between pulses of the heart's magnetic field is modulated by different emotional states, we have performed several studies that show the magnetic signals generated by the heart have the capacity to affect individuals around us.

Biological Encoding of Information

Every cell in our bodies is bathed in an external and internal environment of fluctuating invisible magnetic forces. It has become increasingly apparent that fluctuations in magnetic fields can affect virtually every circuit in biological systems to a greater or lesser degree, depending on the particular biological system and the properties of the magnetic fluctuations. One of the primary ways that signals and messages are encoded and transmitted in physiological systems is in the language of patterns. In the nervous system it is well established that information is encoded in the time intervals between action potentials, or patterns of electrical activity. This also applies to humoral communications in which biologically relevant information also is encoded in the time interval between hormonal pulses. As the heart secretes a number of different hormones with each contraction, there is a hormonal pulse pattern that correlates with heart rhythms. In addition to the encoding of information in the space between nerve impulses and in the intervals between hormonal pulses, it is likely that information also is encoded in the 'interbeat' intervals of the pressure and electromagnetic waves produced by the heart. Low-frequency oscillations generated by the heart and body in the form of afferent neural, hormonal, and electrical patterns are the carriers of emotional information and the higher frequency oscillations found in the EEG reflect the conscious perception and labeling of feelings and emotions. It is proposed that these same rhythmic patterns also can transmit emotional information via the electromagnetic field into the environment, which can be detected by others and processed in the same manner as internally generated signals.

The heart's magnetic field, which is the strongest rhythmic field produced by the human body, not only envelops every cell of the body, but also extends out in all directions into the space around us. The heart's magnetic field can be measured several feet away from the body by sensitive magnetometers. The heart's energy field is an important carrier of information.

Fusion Generators

Fusion power is a proposed form of power generation that would generate electricity by using heat from nuclear fusion reactions. In a fusion process, two lighter atomic nuclei combine to form a heavier nucleus, while releasing energy. Research into fusion reactors began in the 1940s, but as of 2024, no device has reached net power, although net positive reactions have been achieved.

Fusion processes require fuel and a confined environment with sufficient temperature, pressure, and confinement time to create a plasma in which fusion can occur. In stars the most common fuel is hydrogen, and gravity provides extremely long confinement times that reach the conditions needed for fusion energy production. Proposed fusion reactors generally use heavy hydrogen isotopes such as deuterium and tritium which react more easily than protium and produce a helium nucleus and an energized neutron. Most designs aim to heat their fuel to around 100 million kelvins, which presents a major challenge in producing a successful design.

An atom loses its electrons once it is heated past its ionization energy. An ion is the name for the resultant bare nucleus. The result of this ionization is plasma, which is a heated cloud of ions and free electrons that were formerly bound to them. Plasmas are electrically conducting and magnetically controlled because the charges are separated. This is used by several fusion devices to confine the hot particles.

The rate of fusion depends on the temperature and density of the plasma. The plasma loses energy through conduction and radiation. Plasma is an ionized gas that conducts electricity that is modeled using magnetohydrodynamics. Fusion exploits several plasma properties, including:

- Self-organizing plasma conducts electric and magnetic fields. Its motions generate fields that can in turn contain it.
- Diamagnetic plasma can generate its own internal magnetic field. This can reject an externally applied magnetic field, making it diamagnetic.
- Magnetic mirrors can reflect plasma when it moves from a low to high density field.

Superconducting Tokamak

Tokamak is the most well-developed and well-funded approach to nuclear fusion. This method drives hot plasma around in a magnetically confined torus, with an internal current. An advanced superconducting tokamak is often labeled an "artificial sun," that scientists dream will one day create a fusion action in a plasma magnetic containment field that will replicate what they believe happens on the Sun. A Sun on Earth; what could possibly go wrong with that?

After the extraordinarily huge expense of the large hadron collider at CERN, fusion containment is the second largest investment of science in history. So far, the tokamak has only been able to produce a few seconds of fusion reaction. Scientists and engineers are working hard to get hydrogen atoms to fuse together under incredibly high temperatures and pressures, just like they do in the Sun. Artificial fusion power is all about mimicking the Sun's process here on Earth to create a nearly limitless source of energy, or at least that is what nuclear scientists dream about.

Right now, achieving artificial fusion power is one of the most challenging feats in modern science. They have faced tough engineering challenges, including generating extremely high temperatures, stabilizing plasma over extended intervals, and managing precise controls so energy output can be steady.

As of September 2018, an estimated 226 experimental tokamaks were either planned, decommissioned or operating (50) worldwide. One form is called a spherical tokamak, also known as a spherical torus. A variation on the tokamak with a spherical shape. Another form is called the stellarator which uses twisted rings of hot plasma. The stellarator attempts to create a natural twisted plasma path, using external magnets. Stellarators were developed in 1950 and evolved into four designs: Torsatron, Heliotron, Heliac, and Helias.

A tokamak confines a plasma using magnetic fields in a torus. Fusion energy scientists believe that tokamaks are the leading plasma confinement concept for future fusion power plants. In a tokamak, magnetic field coils confine plasma particles to allow the plasma to achieve the conditions necessary for fusion. One set of magnetic coils generates an intense "toroidal" field, directed the long way around the torus. A central solenoid (a magnet that carries electric current) creates a second magnetic field directed along the "poloidal" direction, the short way around the torus. The two field components result in a twisted magnetic field that confines the particles in the plasma. A third set of field coils generates an outer poloidal field that shapes and positions the plasma.

The first tokamak, T-1, began operation in Russia in 1958. Subsequent advances led to the construction of the Tokamak Fusion Test Reactor at Princeton Plasma Physics Laboratory and Joint European Torus in England, both of which achieved record fusion power in the 1990s. These successes motivated 35 nations to collaborate on the superconducting ITER tokamak, which aims to explore the physics of burning plasmas.

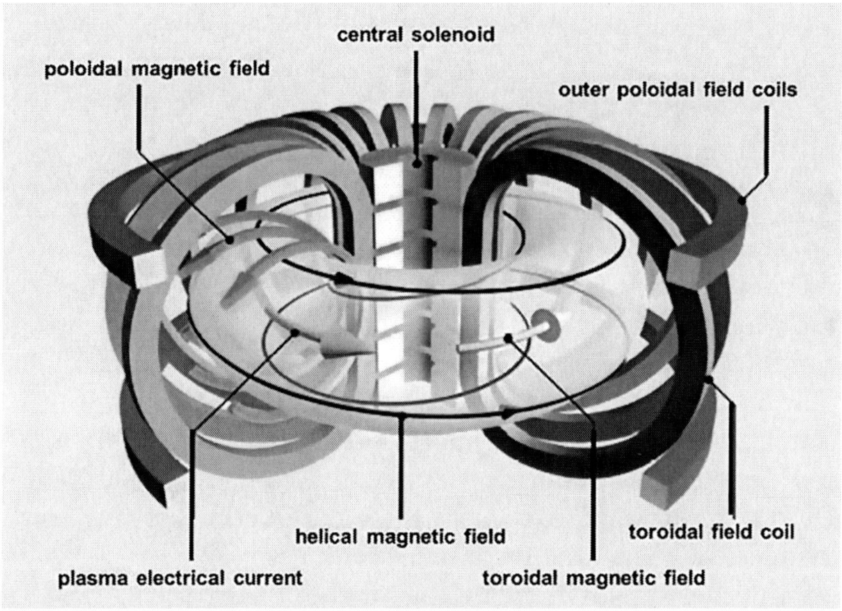

General plan for a tokamak generator

Russia has created a working plasma engine for space travel that utilizes hydrogen as fuel, accelerating charged particles—electrons and protons—to a remarkable speed of 100 km/s (62 miles/s). It is a laboratory prototype of a plasma electric rocket engine based on a magnetic plasma accelerator. Unlike traditional rocket engines that rely on fuel combustion, this propulsion system utilizes a magnetic plasma accelerator and promises to reduce interplanetary travel time significantly. A plasma rocket motor is a type of electric motor. It is based on two electrodes. Charged particles are passed between them, and at the same time a high voltage is applied to the electrodes. As a result, the current creates a magnetic field that pushes the particles out of the engine. Thus, the plasma receives directional motion and creates thrust. In this approach, hydrogen is used as fuel, and the engine accelerates charged particles—electrons and protons.

Failed Attempts at Plasma Containment Fusion Generators

Stellarators create a twisted plasma using external magnets, while tokamaks do so using a current induced in the plasma. Several classes of designs provide this twist using conductors inside the plasma. Early calculations showed that collisions between plasma and the supports for the conductors would remove energy faster than fusion reactions could replace it. Modern variations, including the Levitated Dipole Experiment, use a solid superconducting torus that is magnetically levitated inside the reactor chamber.

Another version is called the magnetic mirror which was developed at Lawrence Livermore National Laboratory in the 1960s. Magnetic mirrors reflect plasma back and forth in a line. Variations include the tandem mirror, the magnetic bottle, and the biconic cusp. A series of mirror machines were built by the US government in the 1970s and 1980s. However, calculations in the 1970s estimated it was unlikely these would ever be commercially useful.

A number of magnetic mirrors are arranged end-to-end in a toroidal ring to create a bumpy torus. Any fuel ions that leak out of one are confined in a neighboring mirror, permitting the plasma pressure to be raised arbitrarily high without loss. An experimental facility, the ELMO Bumpy Torus or EBT was built and tested at Oak Ridge National Laboratory in the 1970s.

The Field-reversed Configuration is a device that traps plasma in a self-organized quasi-stable structure where the particle motion makes an internal magnetic field which then traps itself.

The Spheromak is similar to a field-reversed configuration, a semi-stable plasma structure made by using the plasmas' self-generated magnetic field. A spheromak has both toroidal and poloidal fields, while a field-reversed configuration has no toroidal field. A dynomak is a spheromak that is formed and sustained using continuous magnetic flux injection.

A reversed field pinch moves plasma inside a ring. It has an internal magnetic field. Moving out from the center of this ring, the magnetic field reverses direction. In a z-pinch configuration, a current travels in the z-direction through the plasma. The current generates a magnetic field that compresses the plasma. Pinches were the first method for human-made controlled fusion. The largest such machine, the UK's ZETA, was the last major experiment of the sort. The problems in z-pinch led to the tokamak design. The dense plasma focus is a possibly superior variation.

A theta-pinch produces a current that circles around the outside of a plasma column, in the theta direction. This induces a magnetic field running down the center of the plasma, as opposed to around it. The early theta-pinch device Scylla was the first to conclusively demonstrate fusion, but later work demonstrated it had inherent limits that made it uninteresting for power production.

Magnetized target fusion confines hot plasma using a magnetic field and squeezes it using inertia.

Neutral beam injection ionizing hydrogen which is accelerated by an electric field to form a charged beam that is shone through a source of neutral hydrogen gas towards the plasma which itself is ionized and contained by a magnetic field.

Radio frequency heating uses a radio wave that causes plasma to oscillate (i.e., microwave oven). This is also known as electron cyclotron resonance heating, using for example gyrotrons, or dielectric heating.

Magnetic reconnection uses the principle that when plasma gets dense, its electromagnetic properties can change, which can lead to magnetic reconnection. Reconnection helps fusion because it instantly dumps energy into a plasma, heating it quickly.

Magnetic oscillations use varying electric currents that can be supplied to magnetic coils that heat plasma confined within a magnetic wall.

Inertial confinement is the use of rapid implosion to heat and confine plasma. A shell surrounding the fuel is imploded using a direct laser blast (direct drive), a secondary x-ray blast (indirect drive), or heavy beams.

The Ethers, Plasma, and the Tree of Life

"Then the angel showed me the river of the water of life, bright as crystal, flowing from the throne of God and of the Lamb through the middle of the street of the city; also, on either side of the river, the Tree of Life with its twelve kinds of fruit, yielding its fruit each month. The leaves of the tree were for the healing of the nations."

Revelation 22:1-21

Eastern traditions speak of a land where the aspirant can directly encounter the "rarified" ethers that are only referred to in the highest tantric teachings. This "land" is spoken of in many fashions and throughout almost all mythologies, religions, and spiritual teachings. This land is often called, Shamballa, Tushita Heaven, Hyperborea, Thule, New Jerusalem, or Eden Regained. In this land, the higher ethers of the Tree of Life are available as nourishment for those who are invited into this sacred domain, the holy of holies. Great Masters are found in this realm, and the fortunate initiate is asked to drink heavenly nectar and eat divine ambrosia. These spiritual foods are the gifts of the higher ethers that nourish the etheric and physical bodies of the aspirant as a ‘heavenly dew’ that blesses the efforts of the aspirant. Reaching this land is the goal of the path of self-development.

Of the Tree of Life, man shall not eat, and the Spirit of Matter he shall not hear! These are the regions which are closed to humans—just as the Garden of Eden is now closed to humans. Only through a certain procedure in the old mysteries were the tones of the Music of the Spheres, that brings the cosmic life that is pulsing through the universe, shared with those who were to be initiated. Then, the initiate’s soul could belong to the realms of the Music of the Spheres and to the region of the Word: the living Cosmic Ether.

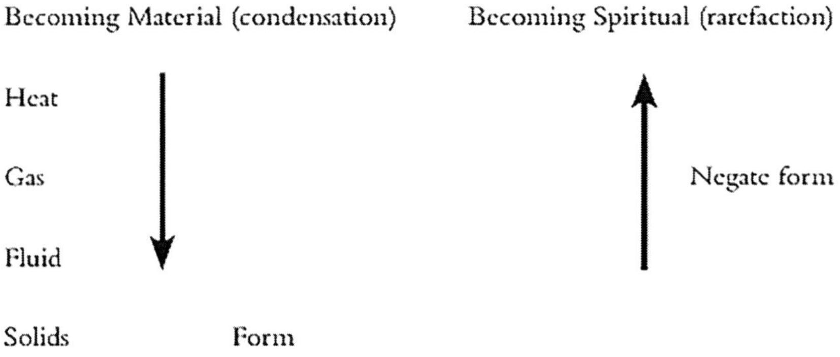

Akasha ether (Mother of All) Future Vulcan

Unfallen Life ether (Tree of Life) Future Venus
Unfallen Chemical ether (Tree of Life) Future Jupiter

Life ether—solid (Earth)
Sound/Chemical ether—fluid (Old Moon)
Light ether—gas (Old Sun)
Warmth ether—heat (Old Saturn)

The phenomenal world of ponderable substance is a medium state extending between the force-activities of the etheric on one side and electricity on the other. The state of substance at any time is the result of this conflict between forces over the mutable world of substance.

The permeation of Warmth and Light ether, the luciferic forces into the earth-sphere, signified for the human organism acceptance of the forces of the "Tree of Knowledge." This event is shown in the *Old Testament* as the "temptation," the eating of the fruit of the "Tree of Knowledge." It indicates that humanity has accepted the forces that were surging in upon him and that he was unable to resist.

The human being could not remain as he was after he had taken the luciferic influences into himself, and thus he needs to be protected from the luciferic influences in his astral body of desires. This was achieved by making the human being of that time incapable of using the whole of his etheric life body. Part of the etheric life body was withdrawn from humanity's jurisdiction. If this beneficial action had not taken place and he had retained power over the whole of his etheric life body, he would never have been able to rightly find his way through Earth evolution. That is the inner meaning of the teaching that humanity had attained the distinction between good and evil.

The partaking of the "Tree of Life" was withdrawn from humanity and they had to exit paradise. That is another way to say that humans retained absolute power over the forces of Warmth ether and Light ether only; power over the formative forces of Sound ether and Life ether were withdrawn from him as the Tree of Life.

Both the sound (Chemical) and life ethers must go through future stages of evolution before they can manifest to human beings as forces that bring life and not death. Humans must evolve further, through the stages of Future Jupiter and

Future Venus, before the two "higher ethers" of the Tree of Life are available for human development. When these ethers are made available, they become a new form of nourishment for humans who have evolved into angels and archangels. These ethers will become the "food of the gods" in a literal way, and the true forces of the Sun will reveal themselves as the Tree of Life.

The Four Ethers

The Tree of Knowledge of Good and Evil—Levity

Warmth Ether—fire ether, chaotizing ether, primal beginnings, Thrones first offering, human consciousness, soul realm, waking consciousness, willing, thermal action, salamanders, ego, red circle forms

The Tree of Life—Gravity

Sound Ether—chemical ether, number ether, form ether, tone ether, harmony of the spheres, folk spirits, inspiration, feeling, dreamless sleep, plants, valence, magnetism, Ahriman, undines, fluids, organs, glands, archangels, sub-atomic, hydraulics, blue semi-circular forms

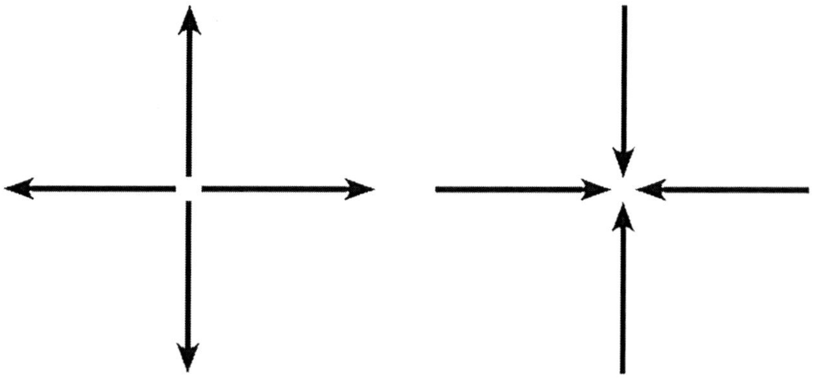

Light Ether—radiance, illumination, imagination, angels, Lucifer, phosphorous, nerves, ATP (adenosine triphosphate), astral, fallen astral, dream consciousness, animals, sylphs, air, pneumatics, electricity, yellow triangle form

Life Ether—word ether, spoken creation, limbs, archai, intuition, time spirits, trance consciousness, minerals, meaning, willing, physical, asuras, radioactivity, gnomes atomic ether, gravity, the third force, solids, fusion, red square forms

At the beginning of our presentation, we stated that it is impossible to speak of the ethers without also speaking of the genesis of space and time and the beings, forces, elementals, and elements that are associated with the etheric formative forces and their rhythmic cycles. To address the first ether that was created demands an understanding of a cosmology that can "hold" the metamorphosing nature of that ether from the most ancient times to the present, and far into the future. Ethers evolve alongside human and spiritual hierarchical evolution. Ethers are not only the building blocks of space and time, but also of consciousness and the beings and forces that work together to create the confluence of Earthly and cosmic (plasma) forces that are found in and around each human.

To the Warmth ether	Leo, Sagittarius, Aries
To the Light ether	Libra, Aquarius, Gemini
To the Chemical ether	Scorpio, Pisces, Cancer
To the Life ether	Capricorn, Taurus, Virgo

Both the world within and the world without are populated by invisible beings whose growth and activity create elements and elementals as a "shadow" of those spirits. Spirits "fall" into matter, and we can witness the forces, effects, and outer manifestation of these beings as the substance of our environment, and likewise our consciousness. Spirit and matter are hyper-connected to the cosmos and the spiritual hierarchy through the ethers. Truly, ethers are magical in their nature and effect.

It is hard to find a place to begin describing the etheric body of a human being. According to many theorists, the ethers pervade all space and all living aspects of the human being. Every atom, molecule, and cell is filled with the dynamic, rhythmic exchange of the four ethers. This pulsing of cosmic and human life is linked, and we can find the duality of the physical plane explicated through the opposing sets of two ethers that alternately give and take in a dynamic field of forces involving electricity, magnetism, levity, and gravity.

Beyond these moving and undulating forces, beings, and substances is the mysterious Cosmic ether, or the Akasha ether. In ancient texts, this primal entity is clearly a being who is the director of the four basic ethers, like a Mother with four rowdy boys who like to wrestle as they express themselves. This Akasha ether is called Fohat by H. P. Blavatsky and others. It assumes the role of the prime creator and prime mover. Akasha weaves between the other four ethers, creating a space for them to play. After hearing all the different theorists, it might be fair to assume that Akasha is primal space that is beyond the spiritual hierarchy and is a donation directly from the Divine Trinity. New theories of the Akasha ether align with the most ancient ones. Akasha is the void of space that all manifestation comes to exist within. It is an ocean of ether, not unlike the mysterious "substance" ether that the ancients referred to with awe. Even Einstein believed in the "hypothetical ether" that was the foundation of his theory of General Relativity, which was his best version of a Theory of Everything. Einstein only referred to this Luminiferous ether quite obliquely, but his concepts of the propagation of light are based upon this ether. Most Theories of Everything have a mysterious force, fluid, substance, vacuum, vortex, or medium in which the forces of the universe act. Many new theories place the ether or ethers at the core of the cosmos. You can find numerous other ether theories that point in the direction of what we have come to see is a complex but understandable system of seven ethers that have different effects on the Earth compared to the cosmos.

The key factors in understanding the ethers is to know how many there are and how they came into being as a process of metamorphoses that evolves over the seven incarnations of the Earth. Not only can the ethers tell us about the past and the creation of the world, but the three "extra" ethers that are added onto the standard four ethers can inform us about the nature of the future. The ethers are a window into viewing the entire plan of human and world evolution. Each ether is connected to a soul or spiritual capacity of the human being. The future spiritual capacities that the human being will develop are also written into the Akasha ether, the Unfallen Sound ether, and the Unfallen Life ether. These three ethers will become more and more available to human use, and this will lead to the future capacities of Imagination in the Spirit Self, Inspiration in the Life Spirit, and Intuition in the Spirit Human. These steps of evolution will take long periods of time to unfold but are available for the initiate at this time. The future can bring to us our future capacities of spirit through the workings of the etheric formative forces.

The human etheric body is an image of the entire cosmos, from ego birth to ego ascension. Thus, the etheric body is virtually immortal and has the past, present, and future inscribed into it as the key to human spiritual metamorphosis. The ethers are the ladder to the spiritual world, and at the first stage of ascension, the initiate gains "Etheric Vision" as an outward sign of development. New, supersensible, spiritual organs of perception develop through the interaction with etheric vision as it is unfolding. Christ is perceived in the etheric realm with this new etheric vision, and He teaches the initiate directly through offering His perfected etheric body as a model to emulate and embody. The higher ego of the individual is the perfected etheric body that exists in the realm where Christ is found: the etheric realm of the Life Spirit. The Earthly ego weds the higher ego, and the human being takes the Akasha ether and consciously weds it to the unfallen sound ether. This brings the active world of Imaginations (Akasha ether) into the world of Inspiration (Unfallen Sound ether—the Harmony of the Spheres). Unfallen sound ether is the Water of Life and part of the Tree of Life. When the unfallen life ether is attained, the initiate merges with the Cosmic Ego of Christ and becomes a co-creator through Intuition. Intuition brings us to the Tree of Life and the cosmic Word made manifest.

The evolution of taking on warmth ether (minerals), light ether (plants), and fallen sound or chemical ether (animals) created the kingdoms of nature that we exist within. These ethers also provide the substance and processes of the realms of consciousness we call dreamless sleep, dreaming, and waking. Thus, the outer world of the kingdoms of nature and the three states of consciousness generally found in the human being, are connected and work from the same principles and with the same ethers. Working with the fallen life ether develops the capacities to recognize eternal truths in ourselves and the world. Gleaning these objective realities leads us over to communion with the spiritual beings who stand behind the outer world of senses.

Fallen life ether is the place where the eternal "I" (or ego identity) of the individual is born. Fallen life ether builds up the human constitution at this time in evolution, but in the future, humans will be able to access unfallen sound and unfallen life ethers. As long as humans are "on the earth" and in the "atmosphere" we experience now, the higher ethers are as yet unavailable, generally speaking. Again, an initiate can rise up to these ethers and pull down the forces and processes of the higher ethers from the future.

While an "earthman," the human being must evolve slowly and with great effort. Suffering time and space is the current fate of human beings unless awareness of the three higher ethers takes place. The gifts of the past are offered through awareness of the evolution of the ethers, and these lessons of humanity are written into the ether body of each individual. The past is summarized in the present etheric body, and the future is insinuated through the evolution of the ethers.

Planet	Ether
Saturn	Warmth ether
Jupiter	Light ether
Mars	Chemical ether
Sun	Life ether
Mercury	Warmth ether
Venus	Light ether
Moon	Chemical ether
Earth	Life ether

The etheric body is the memory body of humanity and the schemata of the future human being. We need to study this magical etheric body and unlock its secrets so that we can share this universal process of metamorphosis as it is found in the world ethers and the human etheric body. We need to link our consciousness to the forces of natural law that we perceive to be manifesting the forces of the ethers.

As we do this, we train ourselves to listen to the etheric body's lessons of the past and begin to imagine the nature and characteristics of future humanity.

As we work with *The Eternal Ethers: A Theory of Everything* as spiritually aware individuals, we can take the ethers, past and future, as the guideposts for human physical and spiritual evolution, both individually and collectively. We can examine the etheric body while it is the focus of human development, from ages seven to fourteen, when the etheric body is most active and building the connectivity to the world and cosmos. It is these years that create a "sensitive period" for an educator to observe the growing child and thus observe the developing etheric body. It is a necessity of our times to understand this development and feed this etheric process the appropriate living images that can help build the individual child's etheric body in a healthy way. It is through images, symbols, and literature that we can "paint" pictures for the growing child to identify with at the particular stage of collective human development that matches their own personal stage of development. The seven-year-old child is like a freshly awakened ancient soul that is trying to come into the modern age over the course of seven years. Each sensitive period or stage of development can be met with cultural offerings that meet the longings of the young child (ancient soul). Each grade, from first to seventh, can be structured to offer an experience of what it meant to be a person in ancient India, ancient Persia, or other cultures that created steps of critical intellectual development that led towards modern-day thinking. In this way, the development of the child replicates the development of the natural human intellect and its spiritual capacities. We have previously pointed at an archetypal system of seven stages that replicate this "memory picture" that we can find in the human etheric body. This scheme is true in the large and the small. The entire cosmos grows through the same stages as a butterfly does, or as a child does from ages seven to fourteen. Part of the process recapitulates the past, while part indicates the future. A teacher's job is to offer soul nourishment for the child's developmental stage through appropriate literature and activities that feed the soul's need to remember and re-embody the wisdom of the past.

Ultimately, understanding the seven ethers constitutes a Theory of Everything. Of course, the best part is yet to come. We can teach the child the past through age-appropriate education up to the seventh grade. In eighth grade, we can help the child become a "citizen of their own time" through summarizing the content of grades one through seven in an objective way. This prepares the child to come into the modern age with a strong association with the stages of human development that have happened over millennia. This is the proper way to welcome and educate the child's etheric body and to build it out of the ethers that have accompanied this evolution. Educating the etheric body is the intent of the Waldorf curriculum. Rudolf Steiner was very conscious of this fact and provided us with the insight to provide every wisdom child with the content of the etheric body: *The Eternal Curriculum.* There are other curricula that can also point at these ideas, but the Waldorf curriculum is the most comprehensive and effective means of linking child development with human development.

Rudolf Steiner on Plasma

Behold, he cometh with clouds; and every eye shall see him, and they also which pierced him: and all kindreds of the earth shall wail because of him. Even so, Amen.

I am Alpha and Omega, the beginning and the ending, saith the Lord, which is, and which was, and which is to come, the Almighty.
Revelation 1: 7-8

The Mystery of the Trinity, Part 2: *The Mission of the Spirit*, Rudolf Steiner, Lecture I, *Meditation: The Path to Higher Knowledge*, August 20, 1922, Oxford, GA 214

"Modern humanity looks at the beginning of the earth, how *it was formed from a primal plasma in space, how life, man*, and—much like a mirage—*the world of ideas arose out of this primal matter.*"

[Note: Life, space and humanity arose from a primal plasma that created the world of ideas.]

Inner and Outer Evolution, Rudolf Steiner, Lecture VI, August 22, 1904, Graal, GA 91

"Atma as form, in particular, it is called Prana. The general Manas as a property of the particular life is called Kama. Kama is thus Manas, claimed for a particular being, distributed to a particular being. In the plants, which already have a similar kind of seed as the animal, one already sees an astral aura, in the simple ones not. In the higher animal, kama already floods in the astral. The physical nature of the mineral, therefore, is essentially nothing but force action; whereas the physical nature of the plant is the substance of which the cell is formed, swelling matter or *plasma*. This *plasma*, only when imbued with kama, is gifted with the capacity to be flesh. The physical nature of the animal is flesh. *Plasma and flesh: they are basically the same substances.* So, what is the difference between force effects? *In plasma, the substances become by being infused with life;* flesh by being infused with life and kama."

[Note: Plasma is flesh, the same substance.]

Mysteries of the Sun and of the Threefold Man, Rudolf Steiner, Lecture III, August 26, 1918, Dornach, GA 183

"Let us make ourselves from a certain point of view another kind of picture of human evolution. We will here keep before us exceedingly important concepts. Now it can be said that when we go back in the evolution of man we discover—and I have often described this—that in ancient days men had more of the group-soul, and that the individual souls were membered into what was group-soul. You can read about this in various cycles: we can then diagrammatically represent human evolution and say: in olden days there were group-souls and each of these split up (it would appear thus to soul perception but different for the perception of the spirit). But each of these souls clothed itself with a body that here in this figure I indicate with red strokes. (Diagram 10).

"Up to the time of the Pythagorean school this drawing, or something like it, was always made and it was said: look at your body, so far as that is concerned men are separated, each having his own body (that is why the red strokes are isolated). Where the souls are concerned, however, mankind is a unity, since we go back—it is true a long way back—to the group-soul. There we have a unity. If you think away the red, the white will form a unified figure (see diagram.)

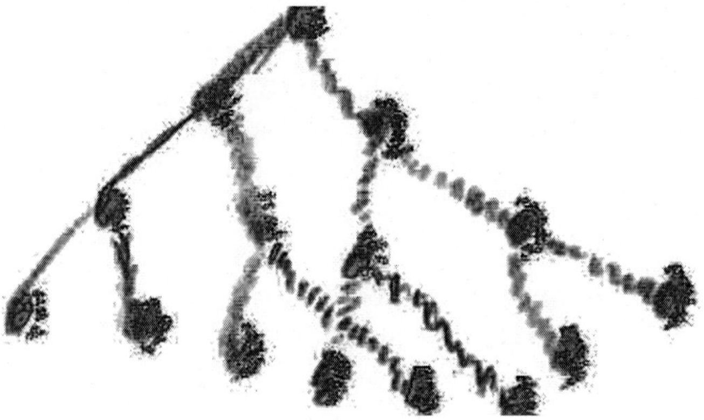

Diagram 10

"There is sense in speaking of this figure only if we have first spoken of the spiritual as has been done here today; for then we know everything that is working together in these souls, how the higher hierarchies are working together on these souls. There is no sense in speaking of this figure if our gaze is not fixed on the hierarchies. It was thus that they spoke up to the time of the Pythagorean School; and it was from the Pythagorean School that Apollonius learned what

I spoke about yesterday and about which I shall be talking further in these next weeks. But then after the eighth pre-Christian century, when the Pythagorean Schools were in their decadence, the possibility of thus speaking was lost. And gradually the concepts that are concrete, that have reality by being related to the higher hierarchies—these concepts have become confused and hazy to people. Thus, there has come to them in the place of Angels, Archangels, Archai, Spirits of Form, Spirits of Movement, Spirits of Wisdom, Thrones, instead of all this concrete weaving of the spirit, they arrived at a concept that now played a certain part in the perception of the Greeks—the concept of the pneuma. Everything became hazily confused: Pneuma, universal spirit, this indistinct concept still so loved today by the Pantheists ... spirit, spirit, spirit ... I have often spoken of how the Pantheists place spirit everywhere; that goes back to Greek life. Again, this figure is portrayed ... but you can now see how what was once concrete, the fullness of the Godhead, now became an abstract concept—Pneuma. The white is Pneuma, the red physical matter (see Diagram 10) if we are considering the evolution of man. The Greeks, however, at least still preserved some perception of this Pneuma, for they always saw something of the aura. Thus, for them, what you can picture in these white branches was always of an auric nature, something really perceptible. There is the great significance of the transition from that constituted Greece to all that was Roman—that the Greeks still in their perception experienced Pneuma as something actual and spiritual, but that the Romans did so no longer. Everything now becomes quite abstract with the Romans, completely abstract; concepts and nothing more. The Romans are the people of abstract concepts.

"My dear friends, nowadays you find in science the same diagram! You can come upon it today in materialistic books on science. You will find the same diagram, exactly the same, as you would have found in the old Mysteries, in the Pythagorean Schools, where everything was still related to the hierarchies. You have it with the Greeks where everything is related to the Pneuma; again today you find it drawn, and we shall see what it has now become. Today the scientist says as he makes this same drawing on the blackboard for his students: in the propagation of the human race the substance of the parents' germ cells passes over to the children; but part of this substance remains so that it can again pass over to the children and again there remains some of this to pass over anew to the children. *And another part of the germ cell substance develops so that it can form the cells of the physical body.* You have exactly the same diagram, only the modern scientist sees in the white (see diagram) *the continuity of the substance of the germ cell.* "He says, if we go back to our old human ancestors and take this germ cell substance of both male and female and then go to present day man and take his, it is still the same stream, the substance is continuous. *There always remains in this germ substance something eternal*—so the scientist imagines—and *only half of the germ plasma goes over into the new body.* The scientist has still the same figure but no longer has the pneuma; the white is now for him the material germ substance—nothing is left of soul and spirit, it is just material substance. You can read this today in scientific books, and it is taken as a great and significant discovery. That is the materialization of a higher spiritual perception that has passed through the process of abstraction; in the midst stands the abstract concept. And it is really amusing that a modern scientist has written a book (for those whose thinking is sound, it is amusing) in which he says right out: what the Greeks still represented as Pneuma is today the continuity of the germ substance. Yes, it is foolish, but today it counts for great wisdom."

[Note: The germ cell is eternal but only half of the germ plasma goes over into the new body.]

The Elemental World and the Future of Mankind, Rudolf Steiner, May 28, 1922, Dornach, GA 212

"What we have considered so far amounts to one aspect of what in the future threatens mankind. There is another aspect—*just as the lower elements of earth, water and air are inhabited by elemental beings, so are the higher elements of light ether, chemical ether and life ether.* However, these beings of the higher elements differ considerably from those of the lower ones. The beings of light, and particularly those of life, do not aim at becoming multitudes. The ones who strive the most to become multitudes are the beings of the earth element. The beings of the etheric element strive rather towards unity. It is difficult to differentiate them from one another; they do not express any individuality and rather strive to amalgamate. Certain initiates in ancient times, through whom certain teachings of the *Old Testament* originated, turned their attention particularly towards the etheric elements. The strong tendency of these elements towards unification created an influence which resulted in the strict monotheism of Judaism.

"The religion which is based on the worship of Jehovah originated mainly from a spiritual vision of the realm of the ethers. In this realm live spiritual beings who do not strive to separate from one another and become many individuals. Rather do they strive to grow together and disappear into one another; they seek to become a unity.

"If these beings are disregarded by man—i.e., if he does not turn to spiritual knowledge and the insight that what exists up in the sky is not merely the physical Sun, but that with the Sun's warmth and light etheric beings stream down to earth; if man's comprehension stops at the external material aspect, then the possibility exists that these beings will unite with ahrimanic powers. In order for the Earth to become what it was originally intended to become, man must wake up to the dangers that threaten from both sides—on the one hand, the danger that those beings who dwell in the lower elements will join forces with ahrimanic powers, and on the other, that the ahrimanic powers will unite with those of the higher elements in their striving for unity.

"The significance of spiritual knowledge for man's earthly destiny cannot be emphasized too strongly. Unless man draws near to spiritual reality, something completely different from what ought to happen will happen to the Earth. No matter how far or how deeply our sophisticated sciences of physics and chemistry investigate the material world around us, the fact remains that what is investigated will all disappear along with Earth existence itself. *In the last resort, chemistry and physics have no value whatever beyond the Earth.* When the evolution of the Earth comes to an end, all mineral substances will turn to dust and dissolve in the cosmos. Only what pertains to the plant, animal and human world will pass over to the [Future] Jupiter existence. Therefore, all the magnificent achievements of these sciences are related only to what is transitory. *It is essential that knowledge is attained concerning that which endures beyond the Earth.*

"As already mentioned, whatever physical laws are discovered, whatever is investigated concerning the atomic weight of individual elements or whatever chemical formulae are produced, all these things are concerned only with what has merely transient significance. Man must grow beyond Earth existence through knowledge of the kind of things I have explained. These are matters of great import and significance."

[Note: Just as the lower elements of earth, water and air are inhabited by elemental beings, so are the higher elements of light ether, chemical ether and life ether.]

Etherisation of the Blood, Rudolf Steiner, October 1, 1911, Basel, GA 130

"*Question:* What is the relation of chemical forces and substances to the spiritual world?

Answer: There are in the world a number of substances which can combine with or separate from each other. What we call chemical action is projected into the physical world from the world of Devachan—the realm of the Harmony of the Spheres. In the combination of two substances according to their atomic weights, we have a reflection of two tones of the Harmony of the Spheres. The chemical affinity between two substances in the physical world is like a reflection from the realm of the Harmony of the Spheres. The numerical ratios in chemistry are an expression of the numerical ratios of the Harmony of the Spheres, which has become dumb and silent owing to the densification of matter. *If one were able to etherealize material substance and to perceive the atomic numbers the inner formative principle thereof, he would be hearing the Harmony of the Spheres.*

We have the physical world, the astral world, the Lower Devachan and the Higher Devachan. If the body is thrust down lower even than the physical world, it comes into the sub-physical world, the lower astral world, the lower or evil Lower Devachan, and the lower or evil Higher Devachan. The evil astral world is the province of Lucifer, the evil Lower Devachan the province of Ahriman, and the evil Higher Devachan the province of the Asuras. When chemical action is driven down beneath the physical plane—into the evil Devachanic world—magnetism arises. When light is thrust down into the sub-material—that is to say, a stage deeper than the material world—electricity arises. If what lives in the Harmony of the Spheres is thrust down farther still, into the province of the Asuras, an even more terrible force—which it will not be possible to keep hidden very much longer—is generated. It can only be hoped that when this force comes to be known—a force we must conceive as being far, far stronger than the most violent electrical discharge—it can only be hoped that before some discoverer gives this force into the hands of humankind, men will no longer have anything immoral left in them.

[Note: Light ether thrust down creates electricity (static), sound/chemical ether thrust down creates electromagnetism, and the Harmony of the Spheres (life ether) is thrust down it creates a terrible force.]

Question: What is electricity?

Answer: Electricity is light in the sub-material state. Light is there compressed to the utmost degree. An inward quality too must be ascribed to light; light is itself at every point in space. Warmth will expand in the three dimensions of space. *In light there is a fourth; it is of fourfold extension—it has the quality of inwardness as a fourth dimension.*

Question: What happens to the Earth's corpse?

Answer: As the residue of the Moon-evolution we have our present Moon which circles around the Earth. Similarly, there will be a residue of the Earth which will circle around [Future] Jupiter. Then these residues will gradually dissolve into the universal ether. On Venus there will no longer be any residue. Venus will manifest, to begin with, as pure Warmth, then it will become Light and then pass over into the spiritual world. The residue left behind by the Earth will be like a corpse. This is a path along which man must not accompany the Earth, for he would thereby be exposed to dreadful torments. But there are Beings who accompany this corpse, since they themselves will by that means develop to a higher stage."

Ethers reflected as sub-physical energies in a sub-natural realm:

- Astral World—province of Lucifer—electro-static–fallen angels
- Lower Devachan—province of Ahriman—electro-magnetic– fallen archangels
- Higher Devachan—province of the Asuras—electro-gravitational—fallen archai

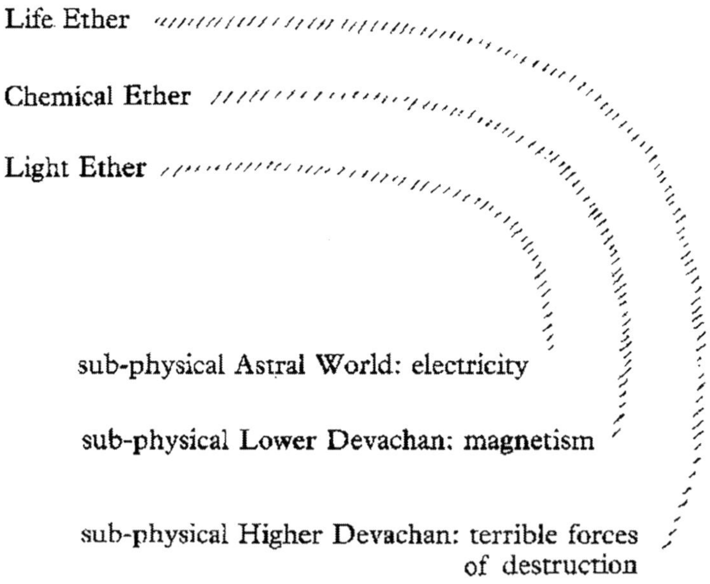

A Road to Self-Knowledge, Rudolf Steiner, *Second Meditation: In which the Attempt is made to form a True Conception of the Elemental or Etheric Body*, GA 16

"The vision of the lightning [plasma] is only something which, as it were, conceals the experience itself; one looks through the lightning to something beyond which is quite different, to something which cannot be experienced in the outer world of sense."

[Note: Plasma cannot necessarily be experienced in the outer world of sense.]

Rhythms in the Cosmos and in the Human Being: Human and Cosmic Breathing — The Light Breathing of the Earth—The Fertilization of Plants and Humans; The Fertilization of Water Through Lightning, Rudolf Steiner, July 20, 1923, Dornach, GA 350

"Now look down at the earth: there, only water rises, not solid parts, as when it becomes a plant; but the water rises. When the plant reaches a certain place up there, it is fertilized by the universe. *When the water that rises here in the form of fog reaches a certain place, it is also fertilized by the universe.* And what happens then? Yes, gentlemen, *there is a flash*! That doesn't always happen, but when fertilization occurs and things are as explicit as they are in summer—otherwise lightning also occurs, but it is invisible—the *water here is fertilized by the universe through light and warmth.* The same thing that happens in the plant happens up there and is visible in the flash. And *when the fog up above has been fertilized, it falls back down as fertile rain.* So, when you see a fog cloud rising, it is actually a—very thin—huge plant;

it opens its blossom up above in the universe, is fertilized, contracts, and the fertilized water droplets fall back down in the rainwater."

"Now you have an explanation for lightning [plasma]. People believe that there is something like a giant Leyden jar or giant electrical device up there; but that is a misconception. In reality, *the water of the earth is fertilized out there so that it can carry out its processes on earth again.* And what happens in the plant happens only much deeper because the plant is more solid; up here near the blossom, when it is the right season, *little sparks happen that you just don't see.* But *these little flashes lead to fertilization.* So, in the appearance of fog and rain, you have the same phenomenon that occurs in the plant during fertilization. And that then goes right up to the human being, where the cosmic breathing, the breathing of the world, which otherwise is only in the head, occurs in the lower body of the human being."

"Then one will say, when one looks at lightning (plasma): Is *lightning* only up there? Oh no, it *is all around us throughout the summer, fertilizing the plants, over the meadows, over the forests, everywhere there is lower lightning.* And in the end *it is lightning that always goes on in us.* Internally we are completely permeated by the same phenomena that we sometimes see when it flashes, and *our thoughts are a flashing within us.* Of course, what once appears as a mighty flash of lightning runs very weakly in our thinking. But now you will also be able to say to yourself: It does make sense to say that *when I look at lightning, the thoughts of the world appear to me, because that is the same as what is in me.* You just have to look at things not superstitiously, but scientifically."

[Note: Water is fertilized by warmth and light from the universe. Lightning (plasma) is all around us throughout the summer, fertilizing plants in a flash of light. Everywhere in summer is “lower lightning” (plasma) all around us and lightning happens in us all the time because our thoughts are lightning (plasma) flashing within us.]

Festivals of the Seasons, Rudolf Steiner, Lecture III, *The Festival of Easter I,* April 10, 1909, Cologne, Various GAs

“For present humanity there stand on the horizon two truths which are intimately connected with one another, two important points for the humanity that is developing and making efforts at the present stage of evolution. These are first, the burning thornbush of Moses (*Exodus* iii) and secondly, the fire that appears on Sinai (*Exodus* xix) amidst lightning [plasma] and thunder and announces to Moses, ‘I AM THE I AM’ (*Exodus* iii, 14). Who is this Being? One who understands the message of Christianity also understands who this Being is who appears to Moses amidst lightning and thunder and gives the ten Commandments. The *Gospel of St. John* tells us that Moses foretold Christ Jesus (*John* v, 45- 46) and, therefore, the power in the burning bush was the Christ. No other Divinity could be imagined when the words were uttered, ‘I AM THE I AM’. He governs invisibly, announcing Himself *beforehand in the fiery lightning upon Sinai.* And he alone understands the *New Testament* who knows that the God announced by Moses is the Christ. The God Who was to bring redemption to mankind made Himself known in a form derived from the fiery element in nature; for therein lives the Christ.

“What is it in the outer world that corresponds to the origin of human thought? Through his thought man experiences a world, and in that form no other being experiences thought. What enkindles thought within us, the simplest as well as the most sublime? Two things co-operate within us when a thought passes through our mind, viz., the astral body and the 'I.' The physical expression of the 'I' is the blood; the physical expression of the astral body is in the nerves. And our thoughts would never flash through us unless there were this interplay of blood and nerves. In the interplay between blood and nerves (inner fire and inner air) we have thought. And this in the cosmos is the rolling thunder when the fiery lightning is enkindled. Imperceptibly to the external world there rings out as thought that which outside is lightning and thunder. *The lightning is like the blood, the thunder like the thought*; and as the lightning gives rise to the thunder, so the *thunder corresponds to the thought produced by the soul.* We see the lightning [plasma] flash out in the air, we hear the thunder, and we feel the blood, we perceive the nervous system, we perceive the thought flash through us and say, 'both are one.' And when the thunder rolls above, it is not a material phenomenon alone; it is actual reality when a person looks up and sees the lightning and hears the thunder and says: *God is now thinking in fire in the way He has to announce Himself to us; this is the invisible God Whose thoughts are in rolling fire*—the God who spoke to Moses. Where would this God have to dwell if He wished to appear on the Earth itself?"

[Note: Lightning (plasma) is like blood, and thunder is like thought and the nervous system. God thinks in rolling fire.]

Lectures to Priests: The Apocalypse, Rudolf Steiner, Lecture XVII, September 21, 1924, Dornach, GA 346

"I would like to give you a few examples of the inner soul language of an initiate, which he doesn't always use outwardly or exoterically, but inwardly in order to form his ideas about his particular experience of the spiritual world. For instance, he might say: Dampen lightning and you will understand color. That is initiattion language. What does it mean? An initiate sees flashes of lightning; he sees something flaming up as it comes out of the universe; he looks upon it as a flickering up of the spirit in world space, and he imagines that when this lightning becomes ever dimmer or milder he gets a dampened or mild development of colors. Lightning spreads out, as it were, and becomes a colored surface. This is the way initiates think of it.

"Or an initiate says: Let thunder become quieter and ever more quiet; then listen to its modulation—and musical things will arise. Thus, initiates look upon the sensory tapestry that is spread out before us as one side of revelation, and it is a real idea for them to think the following: One has the colored manifold world content;—what I'm drawing here could be sounds just as well as colors. It could be like the world content that approaches our senses, that is, like the sensory, physical veil that is spread out before us as our world of perceptions, into which we weave our abstract, unreal thoughts.

"If you imagine that this blackboard is a tapestry that is spread out everywhere and that it consists of sounds, colors and warmth, an initiate sees lightning flashes behind it. The real flashes of lightning that are sometimes visible simply break through this sensory tapestry from behind and from the spiritual world. *Wherever lightning (plasma) appears there is a radiating in of the spiritual world.* And if we look at the way this lightning is dampened and toned down to an even display of colors in nature we have our colorful earth before us."

"If we look up to the heavens and its *stars* (plasma), we have points that also *appear to us out of the spirit, except that they are an enduring, living revelation of flashing lightning*. However, an initiate looks upon all of this as the outer I manifestation of what is behind it. He tells himself:

You should really see—and he does see it, as his soul becomes increasingly active—you should really see the red in a rose begin to spray out above and below as tiny flashes of lightning, and whereas the front part becomes dulled, the red in the rear interacts with the sphere of the Seraphim, and all sounds are linked with the sphere of the Cherubim, and everything that we touch is connected with the sphere of the Thrones. When one sees *nature* around one in the physical world it *is really all an illusion, for in reality it is the toned-down work of the Seraphim, Cherubim and Thrones.*"

"If we look out into the colorful world and at the way it appears,—it is only the evenly toned down lightning flashes and action of the Seraphim. People in ancient times used to speak about the Maya character of the sensory, physical world, but that was only because they didn't know that there are Seraphim, Cherubim and Thrones everywhere in it. Now let's go a little further on the path of initiation. Let's go to where the Apocalyptist emphasizes the opening of the seals. What is going on there? Here color and warmth separate off from the world and increasingly spiritual effects appear which become similar to the true shapes of the lightning-like things which form. Instead of the spiritual world breaking through the sensory tapestry in a zig zag way, we see more gentle lightning flashes, etc., behind it. We know that the beings who live in them are the servants of the Seraphim, Cherubim and Thrones. Something similar applies to sounds, warmth and barely perceptible touches."

"Now just as the earthly, sensory tapestry fades away and such lightning-like and enclosed figures behind it form out of the astral fire and expand ever more, so the stars begin to shine down and their threads of light enable us to see what they are. Starry threads, starry radiations or lights mingle with things that work in an elemental way. Earthly and heavenly things become connected, and we realize that we are in a second world—that was the first one—, and we arrive at the first condition of the second world, where everything is still shining in a natural way, and where we have an inkling that there are beings behind it."

[Note: Lightning (plasma) breaks through into the sensory world radiating in from the spiritual world. Stars (plasmas) appear to us out of the spirit like living revelations of lightning (plasma). Behind the beings who live in lightning flashes are the Seraphim, Cherubim and Thrones.]

***Lectures to the First Class Volume II*, Rudolf Steiner, Lesson XIII, May 17, 1924, Dornach, GA 270**

"In truth it is the spirit from the field of the Cherubim that circulates in the blood that constitutes the voice of conscience. The blood is physical in all the parts of our bodies; but in that it is physical in all the parts of our bodies, it carries the voice of conscience, along with other things. *And the waves of Cherubim life interweave in our blood.*"

"Then we imagine interweaving clouds symbolizing the Thrones. *And in that we imagine these interweaving clouds, we hear the Thrones*, the voices from the first hierarchy: Behold your fiery instinctual drives."

"Now we imagine lightning flashing through the clouds, for *lightning is the tool of the Cherubim, the fiery swords of the Cherubim.* As the lightning flashes through the clouds, we feel these flashes in the words: Behold conscience's soul-guidance."

"Now we imagine the entire sky above the lightning with interweaving warmth, sending the lightning down with heat. And *in this interweaving heat from the cosmos we sense the Seraphim's speech*: Behold your destiny's spiritual trial how destiny extends from earth-life to earth-life to the present earth-life."

"Universal heat; and feel in this threefold "i" [pronounced "ee"] the ascending from the cloud flashes to the lightning bolts *and to the universal heat, from which the lightning bolts derive.*"

"We look up to the lightning bolts. Oh, the Cherubim are already more hidden. We can sense how the Thrones are acting and moving in the clouds. The looming clouds provide the Thrones with substance. The Cherubim do not make it easy for us to see them. They are more hidden than the Thrones. They do not show themselves in the formations. They show us their tools in the lightning bolts. They do not show us their being in the lightning bolts, only their tools."

"And if we *ascend to the cosmic heat, the Seraphim are hidden deeply therein*, much deeper than *the Cherubim are hidden behind their tools, the lightning bolts*. The heat is only the shining of the Seraphim. The Thrones reveal themselves by their being; the Cherubim reveal themselves by their tools; the Seraphim reveal themselves by their shining."

[Note: The Cherubim show us their tools in lightning (plasma). Lightning is the tool of the Cherubim, the fiery swords of the Cherubim. Lightning derives from "universal heat" or "cosmic heat" which is the speech of the Seraphim who reveal themselves by shining. In thunder we hear the Thrones echo lightning and the speech of the Seraphim.]

The Principle of Spiritual Economy, Rudolf Steiner, Lecture VII, *The Macrocosmic and the Microcosmic Fire: Spiritualization of Breath and Blood*, April 10, 1909, Cologne, GA 109

"A thought flashes up in our soul when our blood, our inner fire, nervous system, and air cooperate in such a way. *The origination of the thought inside our soul corresponds to rolling thunder in the cosmos. Likewise, when lightning flashes in the air, and when air and fire interact to produce thunder, this corresponds to the fire of our blood and the activity of our nervous system.* This produces what one might metaphorically call an inner thunder that echoes in our thoughts, albeit gently, quietly, and imperceptibly. The lightning in the clouds corresponds to the fire and warmth in our blood, whereas the air outside, including all the elements it contains in the universe, corresponds to everything that passes through our nervous system. And just as lightning in its counterplay with the elements produces thunder, so the counterplay of blood and nerves produces the thought that flashes up in the soul. Suppose we looked out into the world that surrounds us, saw lightning flashing up in the air, and heard rolling thunder discharging itself. Then suppose that, as we looked into our soul, we sensed an inner warmth pulsating in our blood and felt the life that passes through our nervous system, a thought would flash up within us to tell us that both the external and the internal event were one.

"That is really the truth! Although our thoughts take place within ourselves, the thunder rolling in the sky is not just a physical, material phenomenon. To assert that it was so would be nothing but materialistic mythology. However, individuals able to perceive that spiritual beings weave and radiate through material existence will look up to the sky, see the lightning, hear the thunder, and to them this will be a true and real indication of *God's thinking in the fire, of His intention to announce Himself to us. That is the invisible God who weaves and radiates through the universe. His warmth is in the lightning, His nerves in the air, His thoughts in the rolling thunder; and it was He who spoke to Moses in the burning bramblebush and in the lightning fire of Sinai.*

"The elements fire and air in the macrocosm correspond to blood and nerves in the human microcosm. Thoughts in the human being are what lightning and thunder are in the macrocosm. By analogy, the God whom Moses saw and heard in the burning bush and who spoke to him in the lightning fire of Mt. Sinai appears as the Christ in the blood of Jesus of Nazareth. By thinking like a human being and by being in a human body, Christ's influence as the great model for human evolution extends into the far-distant future. And thus, two poles in the human evolution meet each other: the macrocosmic God on Mt. Sinai who announces Himself in thunder and lightning fire, and the same microcosmic God who is embodied in the human being of Palestine."

[Note: Thought originates in our soul like thunder rolling in the cosmos. When lightning flashes in the air, and when air and fire interact to produce thunder, this corresponds to the fire of our blood and the activity of our nervous system. The lightning in the clouds corresponds to the fire and warmth in our blood, whereas the air outside, including all the elements it contains in the universe, corresponds to everything that passes through our nervous system. And just as lightning in its counterplay with the elements produces thunder, so the counterplay of blood and nerves produces the thought that flashes up in the soul. That is the invisible God who weaves and radiates through the universe. His warmth is in the lightning, His nerves in the air, His thoughts in the rolling thunder; and it was He who spoke to Moses in the burning bramblebush and in the lightning fire of Sinai.]

***The Evolution of the Earth and Man and The Influence of the Stars*, Rudolf Steiner, Lecture XII, September 18, 1924, Dornach, GA 354**

"Let us consider again the explanation I gave of how lightning actually comes about. I told you that *lightning comes out of the overheated air, the overheated universe, the overheated cosmic gas.* I said that there is no question of lightning arising from some sort of friction of the clouds. Clouds, of course, are wet, and if you want to produce miniature lightning with laboratory apparatus, everything must first be wiped absolutely dry. It must not be supposed, therefore, that lightning is a true electrical phenomenon that comes about from the friction of dry elements. It is known that when one rubs glass or sealing wax one produces electricity and so people think that if clouds rub together—well, then there'll be electricity there too. But that is not so. What happens is this: As a consequence of the inner overheating of the cosmic gas, the warmth living in the cosmic gas comes out in the way I have described. Through the fact that the air exerts less pressure toward one side or another, the radiation of the overheated force goes toward that side and lightning flashes.

"Now let us imagine that we have this happening somewhere. In consequence of the greatly overheated cosmic gas—not clouds—the lightning flashes out. And it is quite correct to think that it should stream out in a straight line.

"But you see, it is like this. Picture to yourselves: If an accumulation of heat is present somewhere, it is generally not alone; there are similar accumulations in the neighborhood. In fact, if the Earth is here, let us say, and one looks up there and lightning begins where a concentration of heat exists, then in the neighborhood there are other accumulations: they are not all at one single place. You can imagine, of course, that these accumulations of heat are connected with the Sun's radiations to the various places. Now there are these heat accumulations along the entire path of the lightning and while it is streaming out it snatches up these other accumulations in its course. So, it shines here, then over there, and so on. It takes all the other accumulations with it, and so it moves quite irregularly, and gets this seemingly zigzag formation. The lower it descends, the more it does move in a straight line. There are no longer these heat accumulations; they were higher up. The zigzag of the lightning comes about because it does not arise in one single spot, but from where the heat accumulations are strongest and then it carries the others along on its way. That's similar to when you're out walking and you meet an acquaintance and take him along with you, then the two of you pick up another one, and so on. So that's the story of the lightning."

[Note: Lightning comes out of the overheated air, the overheated universe, the overheated cosmic gas (plasma). As a consequence of the inner overheating of the cosmic gas (plasma), the warmth living in the cosmic gas (plasma) comes out. In consequence of the greatly overheated cosmic gas (plasma) lightning flashes out.]

The Evolution of the Earth and Man and The Influence of the Stars, Rudolf Steiner, Lecture XI, September 13, 1924, Dornach, GA 354

"Lightning is the heat generated in the air itself that makes its way to where there is a kind of hole in the surrounding air, because at that spot the air is thinnest. So, we must say: *Lightning is not caused by electricity, but by the fact that the air is getting rid of, emptying away, its own heat.*"

"Just because of this intensely violent movement, the electric currents that are always present in the air receive a stimulus. *It is the lightning that stimulates electricity; lightning itself is not electricity.*"

"All this shows you that warmth is differently distributed in the air everywhere; this again influences the weather. These are influences that come from the vicinity of the Earth and operate there."

[Note: Lightning is not caused by electricity, but by the fact that the air is getting rid of, emptying away, its own heat. It is the lightning that stimulates electricity; lightning itself is not electricity.]

The Spiritual Hierarchies, Rudolf Steiner, Lecture VII, April 16, 1909, Düsseldorf, GA 110

"These Spirits of Personality have a realm which reaches up to the astronomical Venus (Mercury in the occult sense of the Mysteries). Let us imagine that someone has progressed so far that he is able to observe what is evolving up there on Venus (Occult Mercury). There he can recognize these highly evolved Beings, the Spirits of Personality. When he directs his clairvoyant vision to Venus so as there to observe the assembly of the spirits of Personality, and then sees the lightning flash through the clouds, *he sees in that flash of lightning the reflection of the Spirits of Personality*, for in it they have their physical body."

[Note: Lightning is the reflection of the Spirits of Personality, for in it they have their physical body.]

Genesis, Rudolf Steiner, Lecture VIII, *Stages of Human Development up to the Sixth Day of Creation*, August 24, 1910, Munich, GA 122

"We have spoken of the Cherubim as the directing powers in cloud formation, and of the Seraphim as the directing powers in the lightning flash that issues from the clouds. *If now we look upon the expulsion from Paradise as really referring to a descent from the periphery, we are describing almost word for word how man fell through his own weight, and how he had to leave behind him the forces and the Beings who form the clouds and the lightning—the Cherubim with the flaming sword.* Man falls from the Earth's periphery, out of the region where the Cherubim hold sway with their fiery swords of lightning. There we have a spiritual scientific version that confirms almost word for word the account of the expulsion from Paradise according to which the Godhead placed the Cherubim with the flame of the whirling sword before the gate of Paradise. When you realize this, it becomes almost palpable that those ancient seers who gave us *Genesis* gazed with full powers of seership into the life of man weaving in the etheric heights, before he fell from the regions where the Seraphim and the Cherubim hold sway.

[Note: The Cherubim are the directing powers in cloud formation and the Seraphim are the directing powers in the lightning flash that issues from the clouds. If now we look upon the expulsion from Paradise as really referring to a descent from the periphery, we are describing almost word for word how man fell through his own weight, and how he had to leave behind him the forces and the Beings who form the clouds and the lightning—the Cherubim with the flaming sword. Man falls from the Earth's periphery, out of the region where the Cherubim hold sway with their fiery swords of lightning. The ancient seers who gave us *Genesis* gazed with full powers of seership into the life of humanity weaving in the etheric heights, before he fell from the regions where the Seraphim and the Cherubim hold sway.]

Comet Tails are Plasma

Lectures to the Workmen, The Human Being in Body, Soul and Spirit, Rudolf Steiner, Dornach, Switzerland, April 10, 1920, GA 354

"So, we find meteorites; they are the remains of shattered comets. *But the comets themselves are something quite different; they are alive. The Sun, too, is alive, has a stomach, and not only consumes comets, but eats exactly like we do.*

"Comets are ever-nascent phenomena, perpetually coming into being, passing away again. Out there—outside the system—we need not imagine it to exist at all. It is not there to begin with, but becomes; then, at the perihelion, changes the gesture of its form, which is in fact ever-becoming, ever-nascent. Out there, at last, it melts away again and is no more, *The comet comes into being and passes away*; that is its very nature."

"Altogether, we must look upon the comet as a fleeting thing. In relation to Sun and Earth, *it is a phenomenon of compensation between ponderable and imponderable matter,—a meeting of the two kinds of matter, which do not immediately balance-out* as when light extends in air."

[Note: Comets are alive, and the Sun is alive and eats comets. Comets are always coming into being out of nothing and passing away into non-existence. A comet is a phenomenon of compensation between ponderable and imponderable matter,—a meeting of the two kinds of matter, which do not immediately balance-out.]

The Spiritual Beings in the Heavenly Bodies and in the Kingdoms of Nature, Rudolf Steiner, Lecture VII, April 10, 1912, GA 136

"And in a curious way, the occult vision, which for a time has had the opportunity of observing cometary life, shows us that *everything of a cometary or meteoric nature is ever striving to collect round itself all the harmful astral products, and to remove them out of the planetary system.* We shall see further in the course of these lectures how this affects in particular the harmful astral products of humanity; but we see that the noxious Luciferic evils are carried away through the comets out of the planetary system."

[Note: Everything of a cometary or meteoric nature is ever striving to collect round itself all the harmful astral products, and to remove them out of the planetary system.]

Anthroposophy and Astronomy, by Dr. Elizabeth Vreede, *Anthroposophy a Quarterly Review of Spiritual Science*, Vol. 3, No. 3, Michaelmas, 1928

"We may pursue the same line of study in reference to other cosmic phenomena, the meteors—for they, too, are connected with comets. *Comets are spiritual formations sent forth into the Cosmos by the highest, most sublime spirits, but in the course of their activity as purifiers of the astral atmosphere, they gather to themselves various substances, gases, and the like.* Now other cosmic formations, generated by the 'abnormal' spirits from the realm of the Thrones—Beings who ought really to have become Seraphim and Cherubim but who have remained at the stage of the Thrones—fling themselves in the path of the meteors and work with a prodigious power acquired from their having remained at a lower level. The Thrones are the creators of the mineral group-souls and the retarded Thrones—those who have remained behind—generate solid mineral formations in the cosmos, in the meteors which often accompany the comets or appear in 'showers' at more or less regular intervals of time in the atmosphere of the Earth. And so, in these formations—withdrawn as they are in a certain sense from the ordinary law obtained in the planetary system—we have the highest spirituality united with the densest materiality."

"The very *behavior of comets may continually remind us that the processes taking place in the solar system can never be exhaustively explained by the Newtonian theory of gravitation.* It is only in the comets and meteors that the 'incalculable,' the arbitrary, makes itself manifest. The incommensurable factor in the movements of the other heavenly bodies shows that we must base our observation of their movements on something other than the theory of gravitation which holds good for the Earth."

The Etheric Formative Forces in Cosmos, Earth, and Man

A Path of Investigation into the World of the Living

by Dr Guenther Wachsmuth

Manifesto, by Goethe

You'd study Nature? Then remember
One and All must go together.
Naught is in and naught without,
For what's within is still without.
Hasten man, look up! behold
Her open mysteries unfold!
True her seeming, real her piety;
Rejoice in them and her.
No living thing is one, I say,
Its many, everywhere.

The following is a condensed version of *The Etheric Formative Forces in Cosmos, Earth, and Man*. Rudolf Steiner worked closely with Guenther Wachsmuth on this book, which constitutes the most comprehensive picture of the seven ethers that we find in Anthroposophy. Wachsmuth was Steiner's personal secretary, so we can imagine that he was well-guided by Rudolf Steiner in writing this definitive treatise on the ethers. Below are selections from the book that lay the foundation of what is generally accepted as Steiner's views on the ethers. We present this synopsis of Guenther Wachsmuth's book to contribute to the evolving understanding of the ethers and their relationship to plasma.

There are seven etheric primal or formative forces that are active in the cosmos. However, only four reveal themselves in the space-and-time processes of our present phenomenal world: the warmth ether, the light ether, the chemical ether (or sound ether), and the life ether.

There are seven etheric formative forces, of which four realize themselves in the present phenomenal world:

The Ethers

Akasha ether
Unfallen Life ether (Tree of Life)
Unfallen Chemical ether (Tree of Life)

Life ether—solid
Chemical ether—fluid
Light ether—gas
Warmth ether—heat

The phenomenal world of ponderable substance is *a medium state extending between the force-activities of the etheric on one side and electricity on the other.* The state of substance at any time is the result of *this conflict between forces over the mutable world of substance.*

Akasha

- Life ether [unfallen]
- Chemical ether [unfallen]
- Light ether [unfallen]
- Warmth ether—heat, warmth, fire
- Light ether—gas, air
- Chemical ether—liquid, fluid, water
- Life ether—solid, earth

Man's organism is a microcosm, a copy of the great world organism, the macrocosm. It is through the action of the etheric that life comes to birth. The living cosmic architecture through which these formative forces have produced the evolution of the macrocosm and the genesis and maintenance of the Earth-organism finally reaches a degree of completion in the architecture of the human organism. This not only places and maintains man in a state of harmony with the macrocosmic process, but strives for a higher ascent beyond that he has already reached in the preceding evolution.

The world of spiritual Being streams into the world of substance, bringing to pass involution, evolution, and "creation out of nothing" in an eternal, harmonious circuit. Thus, the cosmic formative forces also stream into the living world of being of the Earth organism, bringing forth, through the manifold reciprocal action of the etheric formative forces, the variations of the world of colors in the world of extension, of space, and of time.

The cosmos must be grasped as a world-organism—the product of active spiritual entities—as a world of organisms whose vital phenomena are the work of the harmonious metamorphosis of etheric formative forces. Cosmos, Earth, and man are built up according to the same will, through the same etheric forces, and of the same substances. The human being who knows himself perceives the being of Nature.

Mother Nature has built him. He builds and masters Nature until human knowledge fully embraces cosmos, Earth, and man.

The living Earth organism has an etheric system of forces in which processes of breathing and circulation take place corresponding to similar life-processes in the human organism. The Earth organism consists of a physical body and an "ether body of the earth," which comprises its etheric formative forces. The "ether body" calls the phenomena of life into existence and regulates and metamorphoses them.

The basis for a new method of studying the etheric will be an organic conception of the various cosmic spheres of etheric action belonging to the planets and heavenly bodies. By including the other attributes of the etheric, we reach a more profound understanding of the "world organism" and its life processes linked with physical motion, etheric currents, and states of consciousness.

The world consists of "states and their transitions one into another," brought about by the etheric formative forces. The world we perceive has been created and is being created by these etheric formative forces. *The spatial comes into existence when the being of a thing passes over into the phenomenal world and mediates the transition from being into phenomenon, and from non-spatial into spatial.*

Through thought, the human spirit seeks to coalesce into a unity what is sundered into multiplicity in the perceptible world. Human thought reverses the action of the etheric formative forces that bring about the phenomenal world; it leads backwards from the phenomenal to real being, and from the separate or spatial into the real or non-spatial. Space and time come to be when the being of a thing passes over into the phenomenal world.

Juxtaposition and sequence are other ways to refer to space and time. The etheric formative forces are the bridge between being and phenomena, but they *are also the bridge between non-spatial and spatial.*

The warmth ether is the primal working entity. When this passes over from the state of rest into activity or, in other words, passes imperceptibly from the "latent" state to our organs and into a state of activity that is perceptible to us, it actually enters into the world of space and time. Gradually, the three additional etheric formative forces proceed from it. These additional etheric formative forces, which weave the perceptible world of color and form, proceed from the warmth ether. This is when, through the action of these four etheric formative forces, the space-time processes that constitute our phenomenal and perceptible world begin.

The warmth and light ethers act expansively—that is, in a space-affirming way—and the chemical and life ethers act contractingly—that is, in a space-denying way. The evolution of the world takes its course, therefore, not merely between primal nebula and heat-death, but between space-genesis and space annihilation—between the affirmation and the denial of space.

First, the space-affirming appeared in the phylogenesis of our cosmic system, and later the space-denying forces appeared. *The genesis of the etheric formative forces is the genesis of space.*

But the *cosmic ether*, from the Earth toward Saturn, is by no means a constant and uniform quantity. This *world ether* is subject—not only as regards space, but also as regards time—to a never-ceasing organic change, though this takes place very gradually, over very long stretches of time. We can divide the cosmos into great fields of space (spheres) that are filled with the etheric active forces of the individual planets.

It is by no means only "electrons" that stream in eternal alternation into the human organism and the living beings of the other kingdoms of nature and out again, but that these are entities that, in their totality, can never be grasped within the range of the theories of modern physics. We do not yet possess any adequate knowledge of phenomena of consciousness associated with the manifestations of force and processes of life and motion in the happenings in the rest of the cosmos.

We perceive that the metamorphosis of states does not merely carry through a meaningless transition from one arbitrary state to some other arbitrary state, but that metamorphosis in the phenomenal world, viewed comprehensively, generally brings an ascent of the evolutionary forms and capacities according to spiritual laws to pass.

The four etheric formative forces have also naturally brought about, in their spheres of activity, the states of aggregation and conditions of substance normal to these forces. The inwarddrawing forces of the chemical and life ethers, tending to concentrate, have worked in toward the center of the earth, produced solid and fluid substances, and assembled these toward the middle spheres of the earth organism. The outward striving, centrifugal forces of the light and warmth ethers have produced the gaseous state of the earth's atmosphere and, in accordance with their tendency to strive outward from the center, have gathered this atmosphere in the outer part of the Earth organism. The etheric formative forces, in their spheres of activity, not only work in the generation of the corresponding states of chemical interrelationships and states of aggregation of substance, *but also in the formation and union of substance.*

The unit of substance consists of four globular spheres that surround one another when they are in a static condition. These may be viewed as the spheres of activity of the four etheric forces, and thus, the great Earth organism—that is, the solid Earth with its atmosphere—is formed so that it consists of four globular concentric spheres, and in each of these four spheres, there is one of the four etheric formative forces.

The suctional, inward-drawing forces of the chemical and life ethers have their principal spheres of activity, corresponding with their tendency, concentrated at the Earth, with its component solid and fluid parts. The outward-tending, centrifugal forces of light ether and warmth ether, on the contrary, have their spheres of activity primarily in the gaseous atmosphere surrounding the globe. This gaseous atmosphere is surrounded by a mantle of warmth ether at the outer borders of the atmosphere, towards the cosmos.

Types of Ether

The four kinds of ether may be classified in two groups. The warmth and light ethers have the tendency to expand and the impulse to radiate out from a given central point. They act centrifugally. The other two—the chemical and life ethers—have the tendency to draw in toward a center, and the impulse to concentrate all in a given central point; their action is suctional or centripetal. This polarity of the two ether groups—the centrifugal, radiating, self-expanding will, and the suctional, centripetal will to draw inward or to concentrate—is an ultimate elemental principle that lies at the bottom of all natural phenomena.

Warmth Ether (Fire)

Warmth ether tends towards the spherical form. Warmth ether is the etheric side of warmth: the inner, impulse-creating warmth, or the warmth of enthusiasm that occurs as the intention that underlies actions. Its elemental counterpart is fire, or externally perceivable warmth. Warmth ether and warmth as an element are closely related. *Warmth is the most rarefied element and is more like a quality than a substance. It represents the transition between the elements and the ethers, and shares features with both.* Warmth as an element can be perceived as the external heat of objects and as body temperature. This warmth is produced when material substances are burned. Warmth as an ether is the inner, impulse-creating warmth that incites activity and that arises when you become enthusiastic about something. It arises within you and has no material features. This is the warmth you need to proceed to action. Because it incites action, warmth ether points towards the future.

Warmth ether is characterized by impulse-creation and directionality. It is a very active quality and is represented by fire as the physical element, which can always burn or grow old. This involves time progressing and has the quality of ripening. In humans, this is found in the will. The fire of love is born from the heart. Warmth appears in time, ripening, and in the human will. *The warmth of blood, or blood as the physical bearer of the I/Ego, is dimensionless, non-spatial, and intensive movement.* Fire begins as a spark, grows to any size, and then shrinks and dies, disappearing from sight. Time progresses like a slow-burning fire. Through an inner glowing or burning, the will of a human is constantly renewing and rising up. Enthusiasm for life is a steady flame that ignites all around it.

Light Ether (Air)

Light ether is invisible. We have learned to recognize other waves that are invisible, like ultra-violet, infra-red, and electric waves, but these have the same characteristics as light waves, differing only in their lengths. Light ether produces triangular forms. It illuminates everything and makes all material things visible. It is also the force that makes plants grow upwards and makes people stand and walk upright. Light ether corresponds to the air element. It makes everything visible by illuminating material things. Light delineates objects and enables you to see their spatial boundaries. You can't see light itself; you can only perceive its presence through the objects and the air it illuminates.

Light radiates from a source. It diminishes with the distance from the source, as it is scattered in space. As the amount of light increases, the illuminated space becomes larger. Light is linear, can be split into two beams by an object placed in its path but can't turn corners. It cannot fill a void around a corner, unlike air.

Light ether induces growth and elongation (plants grow towards the light, while bones in animals and humans grow and elongate under the influence of vitamin D, the production of which is stimulated by light) and makes space expand. Light draws your gaze outward to objects. This shows you that light ether attracts. As seen from within the object (matter), it is a process of elongation; as seen from the periphery (ether), it is a matter of attraction.

Because there is light, there is also darkness. Light ether works between the poles of light and darkness, with the colors of twilight in between. Between light and darkness, there is an area that, on the one hand, has qualities that are mixtures of the two poles (half-light) and, on the other hand, has qualities of its own (such as colors). Light ether is characterized by delineation and the creation of space, linearity and attraction, and polarization. It leads to the periphery and draws the viewer to feel as though he or she is being sucked into that periphery. This also gives the quality of levity, where there is an elongation or uprightness and overcoming of gravity.

Light rays out luminosity, air, growth, and feelings, and it makes things visible and interesting. It stimulates curiosity, admiration, patience, equanimity, and thankfulness. Light is not visible. Imagine a sunrise from darkness to daylight. Slowly, we can distinguish things around us: the objects are seen separated now, space and distances become clear, and space is created through limitation. Light and space belong together. Structure is to see; light divides the inside from the outside and radiates towards the periphery. Light is expanding in space, raying out. The periphery is the constructive principle of light, and the periphery/horizon is sucking our view or consciousness towards the horizon. Light sucks out the sprouts of plants, and in a living organism, it shows as growth. Light brings interest and feelings to the soul that create our relationship to the world around us. Light ether works through air. Air is not visible. It fills the space between objects and has no boundaries. It connects everything. It has no structure or direction and is like chaos because it has elasticity, causing centripetal tension and pressure towards a center. Light ether helps create thinking. It is in strong contrast to matter.

Chemical, Sound, Tone, Number Ether (Liquid)

Chemical ether is also called tone, number, or sound ether. Its forces cause the chemical processes, differentiations, dissolutions, and unions of substances. Its forces transmit the tones perceptible to the senses to us. Space is interpenetrated by waves produced by the forces of the chemical ether, which dissolve and unite substances through geometric, morphological, and harmonic resonance. Chemical ether has a tone and sound nature of which sensible sound, or tone heard by the physical ear, is only an outward expression—that is, an expression that has passed through air as a medium.

Chemical ether produces half-moon forms and is centripetal. Coldness and contraction are ascribed to chemical ether. It is the force that structures the development of phenomena and can be seen in the natural succession of plant communities. It corresponds to the water element. Tone ether is the type of ether that separates and connects. Music is thus based on a force that separates and, at the same time, recombines what was separated in a harmonious way. Tone ether works through proportions, distances, and measures. It separates and recombines, creating a unified whole.

Chemical substances also show certain fixed ratios between the chemical elements themselves, as well as in compounds like salts, acids, and proteins. That's why this type of ether is also known as chemical ether. This ether plays a role in all processes where things are separated and then recombined in new and harmonious proportions. It is characterized by the concepts of separating, structuring, creating proportions, and harmonizing.

The number ether divides, and this creates the numbers (frequencies) and numerical proportions (geometry) of the created world. It brings order and harmony to the growth of all living things through the Harmony of the Spheres and is the basis for all chemical activity between substances, sometimes causing it to be called the valence ether. Chemical ether works through water, which always wants to become a totality: one drop at a time creating mass and integrating like the fusion of two cells during fertilization. Water is the gesture of sympathy; it is thick and heavy; has mass, gravity, and weight; and can be active or passive.

Life Ether (Solid)

Life ether is the most highly evolved ether, and therefore is the most varied and complicated in its qualities. *It is that which is rayed out to us, among other things, from the Sun, and then modified in its action by the atmosphere of the Earth. Life ether, together with chemical ether, belongs to the group of suctional forces: those which tend to draw inwards. It is related to that which is called "gravitation" and to the phenomenon of magnetism.*

Its form-building tendency, when it can exert its effect unhindered in substance, leads to square shapes that are expressed, for instance, in crystallizing salt. It is the force that makes an object, and its environment appear as a unified whole. It is the force that unifies an object's course of life. Life ether corresponds to the earth element and is the life force of organisms. It is the force that creates organic units that are more than the sum of their parts, and in which each component is an integral part of the whole. It is the ether that causes delineation as well as integration. It causes the whole to be visible in each component. It also ensures that each component readjusts to the whole when an organism changes.

Life ether doesn't reject anything and adjusts everything to the whole. It ensures that a changing organism remains recognizable as an individual. It provides continuity and individuality in the biography of an organism. It is also the force that allows an organism to endure and remain itself, thus *maintaining the organism's integrity*. Hence, this ether can restore an organism's health by healing wounds and making them disappear completely in due time.

Life ether is the force that ensures that an organism is a spatially defined indivisible unit, that an organism is and remains itself, even after changes or injury, that an organism is an individual, and that all its components are expressions of the whole and derive their proper place and function from the whole. Life ether affects the individual. *It is the principle that creates integration and individuality.*

Life ether is the great healer. This allows physical processes that have gone "wild" to reintegrate themselves back into the life processes to create the balance of health again. This develops the quality of integrating and uniting. More than the sum of its parts, the life ether brings holistic feelings and flexible thinking. Life ether is individualizing, planar,

square, or cube-forming. It is integrating, healing, flexible, holistic, coherent, and resilient. It coordinates the inside and outside processes of the organs as the parts work together under one idea or a higher aspect. Each organ sends specific forces and frequencies to all the other organs and is balanced by the life ether. We see the harmony of the threefold human being in the nerves, rhythmic system, and metabolism.
Life ether creates the harmony of these three in the mutual interaction of the organs.

Etheric Rhythms

Out of the sublime eternal reciprocal play of day and night activities, chaos and order, summer and winter, near and distant Sun, proceed all the atmospheric and meteorological phenomena. Life actually requires chaotic dissolution: the upsetting of the general laws tending to fixity as substance.

Man, when awake and erect, is coursed through, in his full length, in his brain and vertebral column—that is, in those organs of sense and of thought which, indeed, make him a man—by the Sun current flowing vertically. On the other hand, man, when asleep, has placed himself in such a position that the vertical current from the Sun works only at its minimum upon the brain and vertebral column. Whereas, on the contrary, he is coursed through in his full length when reclining asleep at night by the horizontally acting current from the Moon.

1. The east-west current, a result of the activity of the Sun, exerts its strongest influence in the vertical direction; it stands in reciprocal relation with man, awake, standing erect in waking consciousness, and also to the vital activities of the rest of the kingdoms of Nature during the day.

2. The west-east current—a result of the Moon's activity—exerts its strongest influence in the horizontal direction. It stands in reciprocal relation with the dream-consciousness and deep-sleep consciousness of man, while he is in a reclining position. The animal, on the contrary, on account of the horizontal position of its vertebral column, is exposed, even while awake, to the maximum influence of this current. This current acts at its strongest during the time when the Earth organism, not upset by the Sun, strives to establish its basic structure—therefore, at night. The north-south and south-north currents, have an intimate alternating relationship with soul-life and with the vital activities of the kingdoms of Nature, especially in the course of a year.

Gravity and Magnetism

Gravitation is that which causes substance (bodies) to tend toward the center of the Earth. It *is nothing other than the effect of the suctional activity of life ether centralized in the solid earth.* The magnetism operating from the magnetic pole is only a plus in the sectional pull operating from points where the life ether within the Earth is less disturbed—that is, points of greater concentration. *The magnetic poles* are not fixed spots in the body of the Earth, but *points of the least disturbance, and hence of greater concentration and activity in the force-sphere that belongs to the force of life ether.* As such, these are shifting points—that is, they are centers of force in the etheric Earth.

In our observation of the occurrences within the etheric Earth, we have seen that the etheric Earth—because of rhythmical influences proceeding from the Sun and the Moon—is subject to a constant alternation, expressed in the alternate chaotic upsetting and restoration of the basic structure of Earth. Now, this alternating, disturbing process is naturally at its maximum in the region of the equator, and its minimum in the region of the poles. Accordingly, there must be points within the Earth organism where the constant concentration of life ether and its unmodified action reach their maximum. Such points, then, of the heightened activity of the life ether are the magnetic poles.

Magnetism and gravitation, both on the Earth and in the cosmos, are forms of expression of the same cause—namely, of the life ether. The intensity of the inner tendency of bodies to strive toward a life ether center differs according to their own inner kinship to this ether. *The Sun is a cavity not filled with substance but is a "minus space" or a "vacuity in space,"* as contrasted with the rest of world space, with its substantial and insubstantial contents. Just as "heavy" bodies, more perfectly adapted to the state of life ether, show a greater striving toward the center of the Earth than light bodies, so there is a special form of attraction between the life ether center of the Sun and the magnetic North Pole of the Earth that we have come to know as a point of intensified action of the life ether.

The air in the spheres of the Earth organism are filled with innumerable tiny vacuums (after the manner of the Sun) in which there are suctional forces that, when they issue from these vacuums, cause a drawing together or a condensing of the surrounding substance (water vapor). A condensing process can be established with rhythmic repetitions through which the etheric formative forces cause the genesis of substance, its maintenance, and its reduction—that is, its metamorphosis into another state. We shall understand man—as well as the rest of the kingdoms of nature, with all the phenomena of life—when we acknowledge the living and active etheric formative forces, and the phenomena of matter metamorphosed by these forces within cosmic and earth organisms, in all their manifestations.

Atmospheric Ether

Ozone absorbs life ether with a greater intensity than any other terrestrial substance—that is, it reduces the action of life ether to an intensity that is bearable to the eyes and other organs of man living upon the solid Earth in an atmosphere of twofold oxygen. Pure, unmitigated life ether (in sunlight) would destroy the human eye and the entire human body. The earth organism, through the formation of ozone, prevents the streaming in of direct sunlight, which would be too powerful for his present organs of perception and would destroy him.

Thus, the physical human body can bear the free life ether of the Sun's rays only in the diluted form that has passed through threefold oxygen (ozone) as this diluted life ether, which then appears as the phenomenon light in the twofold oxygen of the lower atmosphere. The layer of ozone in the outermost atmosphere protects man—who lives within the earth organism and in the lower atmosphere—from the intense, destructive rays of the sun. This veil of ozone reduces the intensity of the Sun's rays in such a way that man can perceive the reduced light, which penetrates downwards.

Man, by means of his etheric forces, carries out within himself, in his blood currents and in his breathing, the processes of combustion and of that peculiarly labile transformation of oxygen. Something of the nature of the flame coming into existence under certain conditions and then dying out is also active in man. It is something of the polarity of "heat-light" and "pure light." *The human being bears within himself the secret being of the "flame" of fire,* of which the myth of Prometheus speaks. Light is thus chained to heat in the case of "heat-light," but not in the case of "pure light."

- Pure light requires threefold oxygen ($O3$).
- Pure light is radiated out of the cosmos and into the Earth by the Sun; within the Earth organism, it is given off by phosphorus and by substances that are in process of breaking up (radio-activity).
- Pure light shines as a cold flame. Its action is free from heat.
- Pure light dies under the laws required for heat light.
- Heat light dies under the laws required for pure light.
- Heat light requires twofold oxygen.
- Heat light accompanies terrestrial phenomena of combustion.
- The manifestation of light here is not free but is chained to heat.
- Pure light and heat light constitute an extremely important polarity lying at the foundation of many mysteries within the earth organism that have not yet been unveiled.

Space is interpenetrated by the forces of chemical ether, which differentiate, separate, and unite substance. *Chemical ether is a suctional force, drawing inward; it tends to produce a condensation of substance.* What takes place when the chemical ether, or sound ether, becomes active in the light-ether sphere of the air; the light ether strives for a rarefaction of substance, the chemical ether for a condensation. The tone phenomenon arises as an entity at the moment and at the point at which the conflict begins between chemical ether and light ether over substance, for its rarefaction or condensation.

We are dealing with a polaric contrast. Gaseous bodies represent a surrender of individuality, of shape, and of form; the solid bodies represent a demand for individuality, a thing's own shape, and a special form. Two groups of forces work in polaric opposition: on one side, warmth ether and light ether, the centrifugal forces that tend outward from the solid earth with the Earth's atmosphere; on the other side, chemical ether and life ether, which hold together the Earth organism. It is, therefore, the two latter forces that make the Earth organism possible. Warmth ether may thus be easily demonstrated as a formative force, self-existent and coming to manifestation through the dissolution of substance. A gaseous substance always tends to break through the form imposed upon it; it tends to release itself into the world in every direction. Gas opposes the Earth's laws of form. It will not submit, like the other states of matter, to the earth relationships and tends away from the laws of Earth in order to return to cosmic laws. It denies the sundering, individualizing process toward which Earth tends, as opposed to the cosmos: its tendency is anti-individual.

Water, or fluid, always takes the form of a sphere whose center coincides with the center of the earth. The watery or fluid substance is subject to the laws of the Earth only in part, in that it permits itself to be given its level or its form on its upper surface by the Earth forces tending towards the center of Earth. This shape is common to all fluids in the Earth organism and is, therefore, not individual. For the rest, the fluid has the same form-denying tendency as the gaseous. It is a medium state between purely cosmic and purely terrestrial laws. Solid bodies, finally, take the same form-shaping forces which are common to all watery things in the Earth organism over and into their own interiors; they completely individualize in themselves the "earth state."

The spiral lines described by the several planets in the motions in and with the solar system in the ether ocean of the macrocosm are repeated in the spiral lines that, in different species of plants, describes on the stem as it grows toward the Sun. It is illogical to admit only the action of the Sun and not also that of the planets in the growth phenomena of the plant world. Since the etheric formative forces control the motions of the enormous planetary organisms in the ether ocean of the cosmos just as they control the motions in the micro-organism of the plants, we see in both processes the same rhythms, forms, and motion tendencies.

Etheric Metamorphoses

In the metamorphosis of the ether body lies the cause of change of functions, change of sensitivity to stimuli, change in states of aggregation in cells and tissues, the appearance of new organs and capacities, and the disappearance of others. In the forming of living organisms, we have been able to perceive that the spiritual, the formative, or the substance-creating and form-shaping is the primary creative. This was the case before any beginning and still is. The living organism represents the secondary, the shaped, formed by the formative forces, the object of the creating, the created. The world of substance, woven through by spirit, and the spiritual world working through our phenomenal world is going through an evolution in which we all share.

Just as the Sun cast the planets out from itself in the process of the evolution of the cosmic system and left them behind so that they might follow encircling the Sun, so are the instinct beings and their creations in the phenomenal world—the animals—thrown out. They remain behind in the evolutionary course of the human being, surrounding him at lower stages.

The human being is more highly evolved, and the instinct beings and their organic creations, the animals, are a hardened residue of lower portions of being that the self-evolving human being cast out from himself and left behind in his environment.

We shall generally find in the human organism, in relation to its internal organs, that the right half of the body is more adapted to the forces of the light and warmth ethers, and the left half to those of the life and chemical ethers. Certain organs, even from the earliest fetal stage, incline more to the right half of the body (liver, gall bladder, etc.), and others more to the left half (heart two-thirds left, one-third right; stomach two-thirds left, one-third right from median line; intestines, esophagus, and pancreas tending toward the left; spleen altogether left, etc.). This relationship is highly differentiated between male and female organisms.

If it is principally the life ether working upon a certain part of the organism (the lymph system, the center of gravity, the life ether sphere), then the so-called colorless (white) blood corpuscles appear. If, however, as in the higher layers of the Earth's atmosphere (light ether sphere of the Earth organism), the light ether is chiefly active, red corpuscles appear. The etheric systematization of the blood corpuscle of a man is a copy of the etheric systematization of the interior of the Earth.

The living being, by means of the forces working within him, brings about that labile transformation from twofold to threefold oxygen, and back from threefold to twofold. The blood causes the so-called "slow combustion." This slow combustion is an attribute of all living beings. The blood causes a transitional metamorphosis from the laws of the "warm flame" to those of the "cold flame"; it thus checks the rapid combustion within the organism and renders the slower process of combustion in the living being possible.

Radioactivity exerts a great influence upon life phenomena as ozone and hydrogen peroxide are formed and destroyed. This also happens because of the etheric formative forces active in the process of the circulation of the blood. The organism receives the air—the twofold oxygen—through the lungs and the breathing tissues. It changes this into threefold oxygen ("cold flame," retardation of life process, oxyhemoglobin), and only after a part of the organism has been coursed through in this way does the re-formation take place from O3 to O2 and hereby become heat which, is characteristic of the heat-processes in organisms. While *light ether, together with life ether, forms threefold oxygen*, it is working according to the laws of the cold flame, or the upper or consciousness pole. On the contrary, *when the chemical ether forms the twofold oxygen, it engenders the life element of the warm flame, or the lower or metabolic pole.* Arterial blood or venous blood, slow combustion or rapid combustion, retardation or acceleration of the course of the life processes, protraction of life in the substance world or hastening of the arrival of death—these alternations work themselves out through this eternal conflict.

The conflict between the two etheric formative forces, light ether and chemical ether, which lies at the basis of the life phenomena of the rhythmic system, leads to many and varied goals. It manifests not only in the rhythm of breathing in the corresponding organs, but also in the inner tone-world of man, which leads to speech. It manifests not only in the rhythm of the blood circulation, but also in the substance-structure of this blood in motion—the most vital of all body-building elements.

In the human rhythmic system (breathing, blood circulation, etc.), the light ether works for the upper or consciousness pole, and the chemical ether for the lower, metabolic or substance pole. The one formative force works for the physical basis of the spiritual life, and the other for the physical basis of the corporeal life.

As long as man lives in the substance world, neither of the two conflicting formative forces is victorious, for the living man establishes and maintains an equipoise until the moment of death. The "life" of man manifests as the maintenance of an equipoise between conflicting polarities. Since all these processes are extremely labile and in constant alternation—gradually passing from one into the other—they can never be completely understood by means of chemical formulae. Distinguished investigators admit that the formulae established for oxyhemoglobin, hemoglobin, etc., do not fit the essence of the reality. By means of chemical formulae, it is possible to signify merely with what substances, approximately, the formative force complexes are working, but the substance structure is in continuous change. This is the distinction between living beings and inanimate things, as we have already seen in connection with the Earth organism; the inanimate hardens according to its general laws and according to its structure (rigor mortis, coagulated blood, etc.), but substance permeated by life, on the contrary, is constantly being disorganized, or reduced to chaos, in its combinations and connections. It is modified and prevented from hardening into fixed, dead laws. In that the etheric formative forces prevent this fixing of substance, they maintain the organism in life and set up the great eternal principle of life: without the ether body, the physical bodies of man, of the Earth, and of the cosmos would change into corpses.

It is significant that the human brain is not, like the rest of the organism, subject to the force of gravity directed toward the center of the earth. Since it is subjected to the upward pressure of the cerebral fluid, it is freed from the attraction of the Earth to the utmost extent. The laws of the upper pole of man stand in opposition to the laws of the Earth organism, as these hold sway at the lower pole in man, where the nutritional material taken from the terrestrial environment is worked up, and where no organ necessary to our waking consciousness is formed. In the head (upward pressure by the cerebral fluid means release from gravitation), according to this point of view, laws which are anti-terrestrial hold sway; in the lower man (digestive system), on the contrary, we find predominantly terrestrial laws. Many etheric centers coincide with the nerve ganglia in the human nerve-sense system, but there are more etheric centers than physical centers. Since the same forces—only under different conditions—have formed both the nerve system and the blood system, the second is, in its formation, almost a copy of the first.

Let us first consider the two most important organs: the brain and the heart. As the brain divides itself into an upper and a lower part (cerebrum and cerebellum), so too does the heart. The right and left halves of the brain are distinct, so that there are four parts in total, just as the heart is divided into four chambers. If we now trace this organic differentiation further, it becomes evident that the brain runs out into twelve pairs of nerves.

The structure of the physical heart—which is, at present, still behind the organ of consciousness in its development—has four parts corresponding to the major parts of the brain, but not the twelve-fold division corresponding to the further process of differentiation. The etheric heart possesses this twelve-fold structure.

The human organism possesses a third nerve system in the so-called "sympathetic nerve system," which has its most important organ in the solar plexus and develops its chief activity in the organs of digestion. This is the nerve system of the lower pole, of the metabolic system, and therefore of that part of the organism that is chiefly related to the Earth organism in the assimilation of the nutritional body-building matter taken from the Earth.

Upper pole, head system, Sun influence	*Middle system, spinal-cord nerve system, Moon influence*	*Lower pole, metabolic system, sympathetic nerve system, earth influence*
Brain, 12 pairs of nerves (12 signs of the zodiac)	Vertebral column, 31 pairs of nerves (phases of the Moon in a month)	Assimilation of the substances of the Earth
Voluntary movements	Involuntary movements	Deep-sleep consciousness
Waking consciousness	Dream consciousness	

Ethers in the Earth organism

At sunrise, when Earth breathes out the etheric formative forces that produce the phenomena of life in the plant world, the life current begins to course through the human organism. Beginning at the feet, the current of etheric formative forces enters the human organism and flows through the human rhythmic system, vitalizing many lesser rhythms while it ascends to the brain. In the evening, around sunset, the etheric current flows back to the lower end of the ether body, at the feet.

The etheric current of the breathing of the earth organism rises in the course of the day from the interior of the Earth and then returns there, daily giving a fresh impulse through its rhythms and formative forces to the phenomena of life in all the kingdoms of Nature. It also courses through man, giving an impulse to the vitalizing circulation of his blood and keeping this in continuous flow. The life current courses upward through the human organism primarily on the right side, and downward primarily on the left. Through this fact, the development of the circulation of the blood can be explained.

During the day, the life current ascends and principally affects the nerve system. During the night, it principally descends into the blood system. This creates the heightened waking consciousness of the day and the suppressed dream and dreamless sleep consciousness of the night.

A man who breathes normally takes 18 breaths in a minute, or (18 x 60) 1,080 in an hour, or (1,080 X 24) 25,920 in a day of 24 hours. This is the same number of years in the great Platonic Year, as the Sun precesses through the twelve signs of the zodiac in reverse until it reaches its original spot.

Ethers in Science, Art, and Religion

To find the well-spring from which science, art, and religion are born, we must try to recognize the common destiny of cosmos, earth, and man, and to establish how man can grow into a free creative being. A sublime decree—a rhythm reigning in space and time, an organic plan—forms the basis of this world of forces, which, once its principles are recognized, gives meaning and order to all single phenomena and reveals a part of the hidden primal plan of those creative forces, whereby the when and where of all evolution were established, and are still maintained, in an enduring world-harmony.

The cosmic genesis, from the point of view of time, presents a process of condensation, whether we follow its evolution materially from the "primal nebula" to the solid mineral earth, or genetically from the purely spiritual to the condensed content of our present sense perception. Similarly, in space, a rhythmic process of condensation is indicated in the plan, or the "rough draft of the cosmos."

We recognize the world harmony of these *formative force waves*, surging with their shaping influence, from the outside in, and again from the inside out. Thus, we obtain a deep and satisfying insight into the plan of creation. From an insight into the plan of the world-structure, we can unriddle the purpose, the composition, and the articulation of the links or the "fibers" in the tapestry of the world.

The individual formative forces lead to the following four conditions of substance:

Warmth ether to a warmth condition,

Light ether to a gaseous condition,

Chemical ether to a fluid condition, and

Life ether to a solid condition.

The two forces of the first group operate in an expanding, centrifugal way, while those of the second group have a contracting, centripetal effect. It is a natural consequence of the formative forces that govern there that Mars, Jupiter, and Saturn, which belong to the outer planet-group, are controlled by warmth ether and light ether, and show much less density than Mercury, Venus, and the Moon, which belong to the inner planet-group and are controlled by the centripetal, condensing forces of chemical ether and life ether. Whether we are studying density, light-intensity, color, or any other property of the planets, it will always become explicable by the Ether Theory, and each phenomenon will be evidence for the all-governing principles and harmony imprinted on the universe and its individual spheres by the formative forces.

Four etheric formative forces give rise to and govern our world-organism.

Condition Produced		
Warmth ether	Expanding	Warmth
Light ether	Centrifugal	Gaseous
Chemical ether	Contracting	Fluid
Life ether	Centripetal	Solid

As mighty waves release ever new undulations and rhythms, the creative forces that are surging inward from cosmic space waken the ever-new spheres of force with exalted rhythms. It shows a macrocosmic genesis in space and time, repeating itself rhythmically and increasing toward the center, through ever-new conditions of existence. We see the force-sphere of each planet governed by its special formative force as follows:

Planet	Ether
Saturn	Warmth ether
Jupiter	Light ether
Mars	Chemical ether
Sun	Life ether
Mercury	Warmth ether
Venus	Light ether
Moon	Chemical ether
Earth	Life ether

As the four etheric formative forces have evolved out of one another in a definite series (warmth, light, chemical, and life ether), so do they govern the arrangement of the planetary system in space with similar succession and harmony.

The waves of the four formative forces surge from the region of Saturn to that of the Sun, and then release more waves, which govern the world-spheres between Sun and Earth. The Sun-sphere is the starting point of new force-waves, which is a repetition of the first rhythm.

When the world-rhythms reach Earth, they do not release a rhythm of the same series of force-spheres (warmth, light, chemical, life ether), but a wave tumbling over itself, so that the force-spheres follow one another in exactly the opposite order (life, chemical, light, warmth ether). This process of reversal makes the outside in and the inside out, and is repeated many times within Earth, toward its center. While this third series of reversed waves dies out within the still purely etheric Earth envelope, the organization of Earth mirrors the harmonies of the rhythmic macrocosm. The law of reversal, of turning upside down, governs this reflection of the microcosmic spheres within the Earth.

When we investigate the effects of these cosmic formative forces on the plant growth of the Earth's surface, we must remember that the growth of plants presents a triple rhythm of expansion and contraction. This threefold rhythm of plant growth is only a continuation of that fourfold rhythm in the alternation of expanding and contracting forces that governs the planetary spheres from Saturn to the Earth's surface. When we can see a breathing rhythm of the cosmos in each alternation of expansion and contraction, we recognize that the four breathing rhythms that flow from the outermost spheres of cosmic formative force right in to the Earth's surface are partially reflected there. They then release three breathing-rhythms of plant-growth in the opposite direction. Thus, seven breathing rhythms of the formative forces take part in the cooperation of cosmos and Earth for the creation of a terrestrial plant. By this natural means, the planetary constellations exercise an important influence on the differentiation of plant-growth. We understand now why the ancient Indian wisdom spoke of the "breath of Brahma," and Christian wisdom of the "breath of God."

If we describe the planets Saturn, Jupiter, and Mars as "outer" planets, and, on the other hand, Mercury, Venus, and the Moon as "inner" planets, another distinction is established between the first *force-wave* that formed the "outer," and the second *force-wave* that formed the "inner" heavenly bodies. While, in the first rhythm, the older group of formative forces (warmth and light ether) predominates, in the second rhythm, which is dying out within the world-organism, the latter group (chemical and life ether) generally dominates. Each individual *force-sphere* of the "outer" planets is modified in its influence by the first group of forces, while every individual of the "inner" spheres is modified—that is, strengthened or weakened in its peculiar properties, by the second group of forces. In the outer planets, taken collectively, warmth and light ether preponderate. In the inner planets, taken collectively, chemical and life ether preponderate.

If we study the formative forces of the separate signs of the Zodiac, four groups, each containing three constellations, appear in such a way that each group is governed by one of the formative forces that, together, give the twelve-fold plan of the Zodiac.

Three sets of constellations are thus allied:

To the Warmth ether	Leo, Sagittarius, Aries
To the Light ether	Libra, Aquarius, Gemini
To the Chemical ether	Scorpio, Pisces, Cancer
To the Life ether	Capricorn, Taurus, Virgo

Professor Emanuel Kayser says in *Textbook of Geology*, "Actually the Moon and the Earth form a double planet which moves round a common center of gravity about the Sun. Originally the two bodies must have formed a coherent mass, from which the Moon only later disengaged itself."

Concerning the Moon's re-approach to the Earth from which it was split off, Professor Joseph Plassmann says in the first volume of his *Astronomy*, "The movement of our own Moon has experienced in thousands of years a slight acceleration, independent of its many periodic inequalities, and after study of the relationships between the Earth and the Moon. The attraction must increase and so quicken the movement of revolution; this actual acceleration exists as well as the apparent one; and even though the system moves in so fine a resisting medium, yet the theory shows that the two bodies are getting nearer and nearer to each other, and in the end must crash together."

These statements coincide exactly, then, with the researches of spiritual knowledge, although modern theory generally postpones the re-entrance of the Moon to a distant future reckoned in millions of years, whereas Rudolf Steiner estimated a period no later than 8,000 years hence for the reuniting of the Moon with the Earth. Though the interval of time occupied in the severing and reuniting of the heavenly bodies may still be under discussion, the Nebula Theory and the view of spiritual science are one in this: the whole of the fixed stars, the wandering stars, and the satellites once belonged to a common, united heavenly body and then separated from this and differentiated themselves, and finally strove toward union again.

The planetary spheres outside the Earth are formed, governed, and systematized by a two-fold etheric wave, and each of such waves hears within it—in space—the same rhythm which has led—in time—from the coming into existence of the warmth ether, on to the light ether, to the chemical ether, and to the life ether. This outer two-layered sphere-world is continued inward into the three-layered world of the Earth-spheres. Between the spheres of the outermost Earth-waves, which form the purely etheric Earth-envelope, and the innermost sphere-world, which forms the inside of Earth, there lies, as a rhythmic adjustment, the atmospheric outer covering of Earth. The elastic, atmospheric covering of Earth acts as a separating and softening mediator, bearing within itself something of both polarities and thus making it possible for man to live in a sphere where the war between cosmic and terrestrial polarities is toned down to the more peaceful skirmishing of atmospheric weather-changes.

Mankind of today is exposed to wind and rain, heat and cold, thunder and lightning, hail and snow, the finely powdered dust of former volcanic lavas from the interior of Earth, and the dust of former cosmic meteorite swarms in our atmosphere. These meteorological occurrences are a softening, mitigating adjustment of above and below, outer and inner, through the intermediary spheres of the three-layered Earth, in which the human of today, with his consciousness and his body-building formative forces and substances, can develop his characteristic being.

We can study the configuration of the etheric body of the Earth, which is a reflection of our own inner being, and we must realize that the etheric forces molding the body of the Earth are the same as those in our blood corpuscles. When a change takes place in the etheric structure of the human blood, this will be reflected in the etheric sphere of Earth, because Earth and the blood corpuscles are formed according to the same laws. We are thus led into a quite different relationship with Earth: a truly Christian knowledge of the spirituality of Earth, or a Gaia-Sophia.

The permeation of warmth and life ether, the luciferic forces into the Earth-sphere signified for the human organism acceptance of the forces of the "Tree of Knowledge." This event is shown in the *Old Testament* as the "temptation," the eating of the fruit of the "Tree of Knowledge." It indicates that man has accepted the forces that were surging in upon him and that he was unable to resist.

Man could not remain as he was after he had taken the luciferic influences into himself; he must be protected from the luciferic influences in his astral body. This was achieved by making the man of that time incapable of using the whole of his etheric body. Part of the etheric body was withdrawn from man's jurisdiction. If this beneficial action had not taken place and he had retained power over the whole of his etheric body, he would never have been able to rightly find his way through Earth evolution. That is the inner meaning of the teaching that man had attained the distinction between good and evil. Then, the partaking of the "Tree of Life" was also withdrawn from him. That is to say, man retained absolute power over the forces of warmth ether and light ether only; power over the formative forces of sound ether and Life ether were withdrawn from him.

Principles Derived from *The Etheric Formative Forces*

(The following are condensed and summarized principles found in the book.)

There are altogether *seven etheric primal forces, formative forces, active in the cosmos*; of these, however, *only four reveal themselves in the space-and-time processes* of our present phenomenal world.

The four etheric formative forces have proceeded phylogenetically one out of another, and proceed now ontogenetically one out of another; and, in reality, warmth ether has been metamorphosed—that is, has evolved into light ether; light ether into chemical ether; chemical ether into life ether.

Warmth ether, from which the other ether forces have evolved, *has in turn come into being out of purely spiritual states outside of time and space.*

The first two, warmth ether and light ether, have the tendency to expand, the impulse to radiate out from a given central point; they act centrifugally; whereas the other two, chemical ether and life ether, have the tendency to draw in toward a center, the impulse to concentrate all in a given central point; their action is suctional, centripetal. This polarity of the two ether groups—the centrifugal, radiating, self-expanding will, and the suctional, centripetal will to draw inward, to concentrate—is an ultimate elemental principle lying at the bottom of all natural phenomena.

Ether is simply not matter.

Light ether, when it can exert its effect unhindered in Nature, also produces there triangular forms, whereas warmth ether produces spherical forms.

We shall never be able to reach a true conception of ether by transferring the forms and laws of motion in matter to the ether itself. If, however, we conceive of ether, or the etheric formative forces, as formative forces void of any quality of substance, as active principles which come to living expression in the phenomenal world only through their active tendencies to definite motions, to shaping definite forms with definite qualities, then this difficulty disappears.

The third ether is chemical ether, or sound ether. Its forces, that is, cause chemical processes, differentiations, dissolutions, and unions of substances; but also through activities in another field, its forces transmit to us the tones perceptible to the ear.

Steiner formulates thus the fundamental perception: "Will is the 'idea' itself conceived as force."

The etheric primal forces, formative forces do not belong immediately to the world of phenomena, and they are, like all forces, imperceptible to the physical senses; they belong to a supersensible set of facts. They are to be understood only when we consciously bear in mind that something real, the very beings of things, comes with these forces to living expression in the phenomenal world. The etheric primal forces cannot be understood merely mechanically, nor simply by the negation of all mechanical characteristics. When, as supersensible active principles, they come to living expression in the phenomenal world, they call forth phenomena which may only partially be considered mechanically.

The etheric formative forces are inseparably linked up with individually spiritual and qualitative characteristics. We must ascribe to them, not only such characteristics as velocity, mass, length, volume, etc., which are measurable and calculable, but also characteristics whose laws in the last analysis can just as little be exhausted through numerical estimates as the characteristics of a living man can be exhausted by a table of constants and a sum of mathematical formulae. We shall be able, to form a conception of them only when we observe and investigate them as such entities.

The inner kinship of these two spheres of action will be clear to us from the phenomenon of Chladni's sound-forms. For it is tone which causes the uniting together, the orders and forms, of substance and bodies of substance. That which the physically audible tone produces then in the dust is happening everywhere in space. Space is interpenetrated by waves produced by the forces of chemical-ether, which in the manner of the Chladni dust figures, dissolve and unite substances. But chemical-ether has in reality: a tone-and-sound nature of which sensible sound, or tone heard by the physical ear, is only an outward expression: that is, an expression which has passed through air as a medium.'

Chaldni plate designs created by tone

It may also be proven that the phenomenon of cold is one of those attributes which are to be ascribed to chemical ether, a fact which is essential for an understanding of the relation between processes of cold and of contraction.

The fourth ether is life ether. It is phylogenetically the most highly evolved ether, and therefore in its qualities most varied and complicated, as we shall later show in connection with the most varied phenomena. It is, as we shall see, that which is rayed out to us, among other things, from the Sun and then modified in its action by the atmosphere of the Earth. Life-ether, together with chemical-ether, belongs to the group of suctional forces, those which tend to draw inwards. We shall also be able to prove its relation to that which is called "gravitation" and to the phenomenon of magnetism.

	Spatial	Form	State
Warmth ether	Expansive	Spherical	Heat (Plasma)
Light ether	Centrifugal	Triangular	Gaseous
Chemical ether	Suctional	Half-moon	Fluid
Life ether	Centripetal	Square	Solid

We can never understand the phenomena of cosmos and Earth, of atmosphere, atmospheric electricity, unless we view the Earth as a great living organism which, like all living organisms, is characterized by a process of breathing, circulation, etc.

Life requires chaotic dissolution, upsetting of the general laws tending to fixity. Just as, in the human and animal organisms, the phenomena of life present in albumen and other organic substances arise from the breaking up and chaotic upsetting of the fixed, normal, and characteristic chemical states basic to them, so also the life of the Earth organism depends upon the continuous breaking up of its primal normal state of conformity to laws through cosmic influences, and especially those of the Sun.

Out of the sublime eternal reciprocal play of day and night activities, chaos and order, summer and winter, near and distant Sun, proceed all the atmospheric and meteorological phenomena. The four etheric formative forces have also brought about naturally in their spheres of activity the states of aggregation and conditions of substance normal to these forces. Thus, the inward-drawing forces of the chemical and life ethers, tending to concentrate, working in toward the center of the Earth, have produced the solid and fluid substances and assembled these toward the middle spheres of the Earth organism; whereas the outward-striving, centrifugal forces of the light and warmth-ether have produced the gaseous state of the Earth's atmosphere and—in accordance with their tendency to strive outward from the center—have in reality gathered this atmosphere in the outer part of the Earth organism.

Barometric pressure comes from the rhythmical tension and release of the suctional centripetal forces of chemical ether proceeding from the solid Earth.

The alternating barometric pressure is due in like manner to this: whether the Earth organism sends its inward; drawing, suctional etheric forces above the solid Earth into the atmosphere—that is, exhales; or draws these back—into their own spheres of activity—that is, inhales.

We shall be able to follow the chemical ether in its rhythmical activity of creating moisture. Both dew and mist in the morning and evening are also the results of the moisture producing activity of chemical ether, because in the morning and evening this reaches its highest state of concentration in the lowest levels of atmosphere at the surface of the Earth. The Earth organism may be said to breathe in and out, along with the chemical ether, the moisture-producing elements in rhythmical alternation.

Chaos is induced by the coming together of the normally separated etheric formative forces.

Electric and etheric currents encircle the globe.

Gravitation causes substances and bodies to tend toward the center of the Earth, is the effect of the suctional activity of life ether in the centralized solid Earth.

Those bodies in which the tendency to become identified with the related life ether center of the Earth we call "heavy bodies." In contrast, bodies whose structure (fluid, gaseous, etc.)—to be ascribed more to the chemical, light, or warmth ether which is less adjusted to the life ether, have correspondingly less tendency, or none at all to strive toward the life ether center of the Earth. These are "light bodies."

The Earth does not rest until it has brought the solid, fluid, gaseous substances within its organism to the spheres of the corresponding formative forces.

Accordingly, there must be points within the Earth organism where the constant concentration of life ether and its unmodified action reach their maximum. Such points of the heightened activity of the life ether are the magnetic poles. It is a result of the action of a truly variable field of force.

Life ether is the uniform common cause of gravitation and terrestrial magnetism.

Considered from the etheric point of view, the Sun is a space not filled with substance, but in which the etheric forces of life ether are active.

Naturally, life ether in small quantities enters also into the atmosphere; indeed, it is this which holds the atmosphere together by its suctional action, since otherwise—in accordance with the tendency of light and warmth ether and the gaseous substances adapted to these—the atmosphere would fly away into cosmic space.

We see principally in discussion of the variability of "gravitation," terrestrial magnetism, the double wave of barometric pressure, the rise and fall of water in the vegetable kingdom, the inhaling and exhaling of chemical ether by the

Earth organism, etc., so that the chemical ether also works in rhythmic alternation within and without the solid Earth.

The difference between the Sun and the inner Earth consists in this, that in the Sun the life ether works only as a free etheric force and therefore induces the tremendous magnetic intensity of the Sun, whereas in the inner Earth it works reciprocally with substance and therewith produces the solid state of aggregation of the fourth sphere (shell of the Earth) adapted to the life ether, or else produces modifications of this through the co-operation of the other formative forces toward the center.

1. Warmth ether 2. Light ether 3. Chemical ether		*Outside of earth* (Atmosphere)
4. Life ether		Solid earth
5. Chemical ether 6. Light ether 7. Warmth ether	Here modified by life ether	*Interior of earth*

Wherein 1=7
2=6
3=5
4

We have found that the etheric formative forces of the Earth organism, the ether body of the Earth, are influenced by two fundamental etheric currents of force in the cosmos; that is, by the life ether streaming to us most of all from the Sun, and the chemical ether radiated to us chiefly from the Moon, with their respective vertical and horizontal directions of currents. The more recent scientific research has now made the exceedingly important discovery that the most abundant substances in the outermost strata of the atmosphere are helium and hydrogen. More- over, we have learned to recognize oxygen as the most important ingredient in the atmosphere, so that we may look upon hydrogen and oxygen as two of the most essential basic elements in the material world. Moreover, our breathing process and our circulatory process are fundamentally dependent upon oxygen and its modifications and upon the watery state of the blood.

For we shall learn that the entity of two-fold oxygen is distinguished always from the entity three-fold oxygen, not only in that oxygen at one time enters into this combination and another time into the other combination but that the difference between these two entities is most intimately connected with the phenomena of life and death. For the most important point here is, not the number of the units of oxygen combined, but much more this question: What forces bring about, on the one hand, the combination $O3$, and what forces the combination $O2$; and what forces are linked, therefore, in their activity with the presence of the substance-structure $O3$ at one time, or the substance-structure $O2$ at another?

If we now clarify our minds as to the ontogenetic origin of substance, we shall find the following to be true of the Earth organism, in which the two etheric formative forces, life ether from the Sun and chemical ether from the moon, are active.

1. Experimental research establishes the fact that the substance which makes its appearance in the vanishing of matter, in its spontaneous radio-active decay, is the same as that which we discover at the boundaries of our Earth-organism, in the coming into existence of matter: that is, helium (He).

2. When the forces radiated inward into the Earth organism penetrate more deeply into the material world, the Sun force then forms ozone, threefold oxygen in reciprocal action with the forces and substances of the light ether sphere of the

> Earth organism. Jr. K. Kahler, of the Potsdam Meteorological Observatory, says in his book "Luftelektrizitat," that there "is thus indicated a mutual dependence between a (the intensity of electricity diffused in the atmosphere) and the ozone content of the air. With the increase of air, the ozone content also becomes greater. We have to do here, at least, with an indirect relation which seems to be attributable to the rays of the Sun." And elsewhere: "Another source of electricity, chiefly important for the uppermost strata of the atmosphere is the ultra-violet Sunlight. It produces ozone."

On the basis of a concrete understanding of the activity of the etheric formative forces we will express the matter thus: The life ether radiated from the Sun into the Earth organism forms in the light ether sphere ozone, the threefold oxygen.

The important distinction between the oxygen in the threefold state on the one hand and oxygen in the twofold state on the other, will in future play an important role in any phases of the study of organisms.

In the same spheres of the Earth in which the life ether of the Sun forms ozone, or threefold oxygen, the chemical ether (of both the Moon's radiation and also the Earth's radiation) forms hydrogen peroxide, (H202). In view of the intimate association of this substance with processes in the human organism, this also will play a great role in future organic research. Here, we should fix firmly in our minds that these two substances, threefold oxygen and hydrogen peroxide, so intimately connected with the phenomena of life, have come into existence by reason of the penetration of the Sun forces (life ether) and the Moon forces (chemical ether) into the upper light ether sphere of the Earth organism.

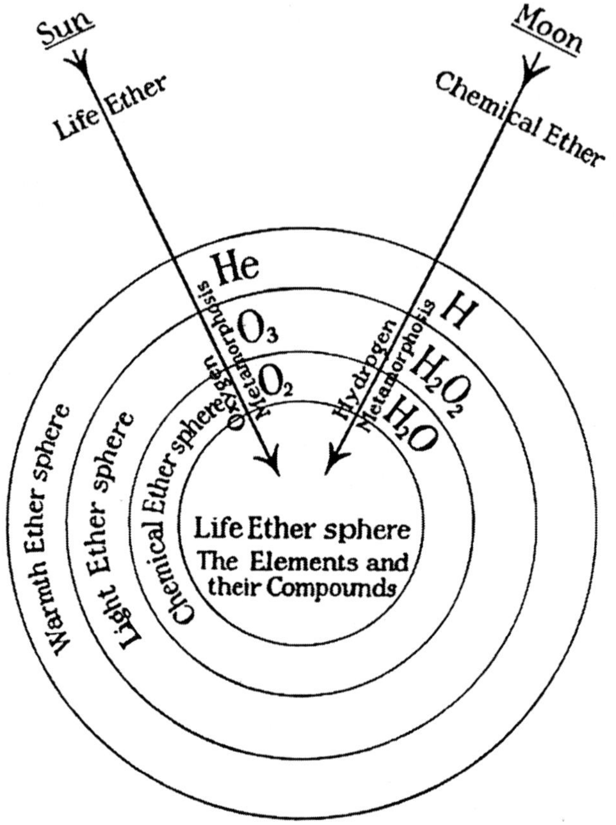

In the case of the formation of water in the clouds, it is the action of chemical ether sucking together the water vapor and thickening it to water. We must here distinguish sharply also between the formation of water and of clouds, on the one hand, where it is due to the action of cosmic forces radiated into the Earth organism (chemical ether of the Moon) and on the other hand such formation of water as results from the radiation of forces outwards from regions of the Earth. We have already shown the mutual connection of the formation of water and clouds and the phenomena of the northern lights, the polar lights, where likewise, it is chemical ether which produced the water, but chemical ether radiated outward from the interior of the Earth.

A condensing process—with rhythmic repetitions—can be established, through which the etheric formative forces cause the genesis of substance, its maintenance, and its reduction, that is, its metamorphosis into another state.

Being ↑		
Mediating Forces	Life ether Chemical ether Light ether	Free etheric formative forces
Phenomenal World	Heat —(Warmth ether) Gaseous —(Light ether) Fluid —(Chem. ether) Solid —(Life ether)	States in the phenomenal world caused by the etheric formative forces.
Mediating Forces	Sub-field (Electricity)	Forces united with substance (sub-material)
Being ↓		

The phenomenal world of ponderable substance is a medium state extending between the force activities of the etheric on one side and electricity on the other. The state of substance at any time is the result of this conflict between forces over the mutable world of substance.

The etheric formative forces are the bridge between being and phenomena, but thereby at the same time the bridge also between non-spatial and spatial. For only when they begin to work does there originate a juxtaposition and a succession that is, space and time.

Pure-light phenomena always give rise to the threefold oxygen (ozone-O3); indeed, it is really bound up with the presence of ozone—that is, it cannot appear or manifest itself to human perception apart from ozone. Heat-light, on the contrary, or the light phenomenon dependent upon the co-operation of warmth ether, is always linked with the twofold oxygen,—that is, it cannot appear or manifest itself to human perception apart from twofold oxygen. Unlike pure-light, it cannot manifest itself in threefold oxygen, ozone. The prerequisite for the existence of "pure light" in the phenomenal world is threefold oxygen; for "heat-light" twofold oxygen. These appear on the one hand in Sunlight—that is, in the rays penetrating into the Earth organism from without; and on the other hand, within the Earth organism itself in the phenomena of phosphorescence and the like, as well as the phenomena of radio-active decay.

1. Cosmic Pure-Light

The Sun's rays of life ether, when they penetrate the outermost layers of the Earth organism, form helium; as they enter further and reach the sphere of light ether, there they form ozone, the threefold oxygen.

The formation of threefold oxygen, or ozone, as a transitional stage in the passage of the Sun's rays toward the solid Earth, is of the utmost importance for rendering Sunlight (pure light) visible to the human eye. For ozone absorbs life ether with a greater intensity than does any other terrestrial substance—that is, it reduces the action of life ether to the intensity which is bearable to the eyes and organs humans living upon the solid Earth in an atmosphere of twofold oxygen.

Pure, unmitigated life ether (Sunlight) would destroy the human eye and the entire human body. Abundance of proof can be adduced by the mitigating action of ozone.

Ozone, in which life ether is always held in absorption, destroys human skin, vegetables, and other organic substances when large quantities of it are brought in direct contact therewith.

The physical human body can bear the free life ether of the Sun's rays only in the diluted form which has passed through threefold oxygen (ozone), as this diluted life ether then appears as the phenomenon of light in the twofold oxygen of the lower atmosphere.

In this fact we have one reason why the light ether sphere is thus designated, since it is in this sphere of the Earth organism that the macrocosmic etheric forces streaming from without are reduced to such less powerful forces and convey the perception of the light phenomena to the human physical organs.

The Earth organism is constituted with wonderful wisdom. By forming a layer of ozone in the outermost atmosphere, it protects man, who lives within the Earth organism in the lower atmosphere, from the too intensive, destructive rays of the Sun, and by means of this veil of ozone reduces the intensity of the Sun's rays in such a way that man can perceive the reduced light which penetrates downwards. This is a fact of fundamental importance for all phenomena of life on the solid Earth.

The man of today, then, if he were living in the upper light ether sphere of the Earth organism, would be an entirely differently constituted being, with quite different organs of perception and a correspondingly different state of consciousness. His organs of perception and his consciousness are now of such a character that, in his waking state, he can perceive light as it is brought to him in the lower strata of the Earth organism, the state of this light being due to the fact that the Earth organism, through the formation of ozone, prevents the streaming in of a different light which would be too powerful for his present organs of perception and would destroy him.

This is certainly the most impressive evidence of the fact that in phosphorus appears the action of the same forces which give rise to cosmic pure-light in the higher light ether sphere of the Earth organism. For its activity in creating light ceases when one arbitrarily subjects phosphorus to the laws of the lowest atmospheric strata. In the realm of cloud-formation (upper light ether sphere of the Earth organism) there are active the same etheric forces which manifest their activities below on the Earth in phosphorus. The flame of phosphorus is cold. The flame of phosphorus is so cool that a person can comfortably hold his hand in it.

Phosphorus shows all the characteristics of "pure-light":

1. It forms ozone,
2. It requires the laws which are characteristic also for cosmic pure-light of the upper atmosphere,
3. Its light is generated without the co-operation of warmth ether, and
4. Its flame is cold.

2. Earthly Heat-Light

Earthly heat-light is a light phenomenon which depends upon the terrestrial process of combustion, or the co-operation of warmth ether. While the phenomena of pure-light, which depend upon the action of life ether and light ether and in whose genesis warmth ether does not co-operate, are still in a very general way free from arbitrary human control at both poles of its manifestation, as Sunlight streaming into the world and also within the Earth organism in connection with radio-activity, phosphorescence, and the like. The other light phenomena (heat-light) which depend for their manifestation upon the co-operation of warmth ether, are completely subject to arbitrary human control.

The theory of ether now enables us to understand how man, by means of his etheric forces, carries out within himself, in his blood currents and in his breathing, the processes of combustion and of that peculiarly labile transformation of oxygen. In man also there is active something of the nature of the flame coming into existence under certain conditions and then dying out, something of the polarity of "heat-light" and "pure-light." Man bears within himself the secret being of the "flame" of fire, of which the myth of Prometheus speaks.

Light is thus chained to heat in the case of "heat-light," but not in the case of "pure-light." The fact that, in the process of combustion accompanied by light, the co-operation of life ether is also essential for the appearance of the phenomenon of luminosity—is shown by the fact that gases (that is, substances which are not found in the life ether state) cannot of themselves produce a luminous flame, whereas on the other hand the solid bodies (that is, substances which belong in the life ether state) send out rays of light when glowing hot. It is therefore obvious that the life ether freed through the destruction of solid substances is necessary to enable the combustion process to become luminous.

This new classification of manifestations of light within the Earth organism as pure-light and heat-light has thus demonstrated its correctness by reference to phenomena; likewise have the marks of their distinction—difference in relationship to one primal substance—for pure-light requires threefold oxygen where heat-light requires twofold.

Pure-light dies under the laws required for heat-light. Heat-light dies under the laws required for pure-light. Pure-light is radiated down out of the cosmos into the Earth by the Sun; within the Earth organism it is given off by phosphorus and by substances which are in process of breaking up (radioactivity). Pure-light shines as a cold flame. Its action is free from heat. Heat-light accompanies terrestrial phenomena of combustion. The manifestation of light is here not free but is chained to heat. Pure-light and heat-light, constitute an extremely important polarity lying at the foundation of many mysteries within the Earth organism which are not yet unveiled. We shall again find these polarities between forces most of all in man and his organism.

Every color is the result of a conflict between the real etheric forces of light and darkness.

Living plants will answer to the presence of a certain color by an expansive or growing process (warmth ether and light ether), while the presence of a polarically opposed color will be accompanied by a dis-uniting or cell-separating process (chemical ether and life ether).

Chemical ether is active not merely in its own manner in the differentiation, the division, and the uniting of substances, but it is the forces of chemical ether also which transmit the tone perceptible to the senses. The inner contact of the two realms becomes clear to us in the case of "Chladni's figures." There, it is tone that causes the weaving together, the arranging and shaping of substances and substance forms. What audible tone affects the sand is also happening everywhere in space. Space is interpenetrated by the forces of chemical ether which differentiate, separate and unite substance. Chemical ether has a tonal sounding nature, of which the audible sound heard by the physical ear is only an external expression—that is, an expression which has passed through the air.

We have shown the "spectrum of Nature" in the following manner:

Life ether
Chemical ether
Light ether
Heat: Warmth ether
Gaseous: Light ether
Fluid: Chemical ether
Solid: Life ether

Let us follow it in its activity in the sphere belonging above it, according to the description previously given, the gaseous state—that is, the surrounding air, the atmosphere. The gaseous state is produced by light ether, an expansive force. It represents a rarefaction of substance. Chemical ether on the contrary, is a suctional force, drawing inward; it tends to produce a condensation of substance. What takes place when the chemical ether or sound ether, becomes active in the light ether sphere of the air? A conflict! The light ether strives for a rarefaction of substance; the chemical ether for a condensation.

Life ether and light ether will form ozone; warmth ether will destroy it; chemical ether will form water; light ether will not.

Like the cosmic formative forces which penetrate into the outer spheres of the Earth organism, life ether not only forms the ozone, so vitally important, but reaches deep into the phenomena of life. The physiological effects of these released forces are almost all-comprehensive.

The model of the Earth organism is also the model of the smallest units of substance [atoms], which, however, go through their metamorphoses according to the different influences of the various regions of the Earth and thus are more and more varied and individualized in space and time.

The life ether as a free force brings about not only the sundering of the Earth organism from the cosmos, its individualization, but also the sundering from the Earth organism of each single body within it.

By taking in the life ether, the force of gravity, it has taken over within itself the Earth force; it is individualized. We may summarize as follows: Gas opposes the Earth's laws of form; will not submit, like the other states of matter, to the Earth relationships; tends away from the laws of the Earth in order to return to cosmic laws. It denies the sundering, individualizing process toward which the Earth tends as opposed to the cosmos; its tendency is anti-individual.

Warmth ether (heat) and light ether (gaseous substances) have anti-individual form tendencies, chemical ether (fluid substances) have partly anti-individual, partly individualizing form tendencies, whereas life ether (solid substances) have purely individualizing form tendencies.

The Senses and the Ethers

The sense of heat and warmth ether

The sense of sight and light ether

The sense of taste and chemical ether

The sense of hearing and chemical ether

The sense of smell and life ether

It is illogical to admit only the action of the Sun and not also that of the planets in the growth phenomena of the plant world; for, since the etheric formative forces control the motions of the enormous planetary organisms in the *ether ocean of the cosmos* just as they control the motions in the micro-organism of the plants, so we see in both processes the same rhythms, forms, and motion tendencies.

The predisposition of the species cell—that is, that which is inherited in the embryo, which determines by law what it will become—is the ether body. The essential, therefore, in heredity is not the substance of the germ cell but its etheric structure—that is, those etheric formative forces which chiefly determine the coming into existence of this organism.

We have already shown that, if one wishes to understand the structure of the body of Saturn as this has been established by modern astronomy, one must take as a foundation the fact that the etheric structure of the more highly evolved Earth represents a complete reversal of the etheric structure of Saturn: that, therefore, what is inner in the body of Saturn has become the outer in the body of the Earth, and the outer in the body of Saturn has become the inner in the Earth. Moreover, we have set forth that this reversal process repeats itself in the organism of the Earth itself, in that the etheric structure of the interior of the Earth represents a reversal of the etheric structure of the exterior. We perceive, then, that this mighty macrocosmic process of differentiation, which changes the outer to the inner through the reversal of the etheric structure and thereby alters and furthers the evolutionary conditions of each tremendous planetary organism, also reappears in the evolutionary process of the embryo as the most important factor in its evolution. What has brought to pass the genesis of the macro-organism, this brings about also the genesis of the embryo. The coming into being of the organisms proceeds both macrocosmically and microcosmically according to the same harmonious laws; both there and here it is the work of the etheric formative forces.

Physical Body and Basic Structure of the Ether Body of Man

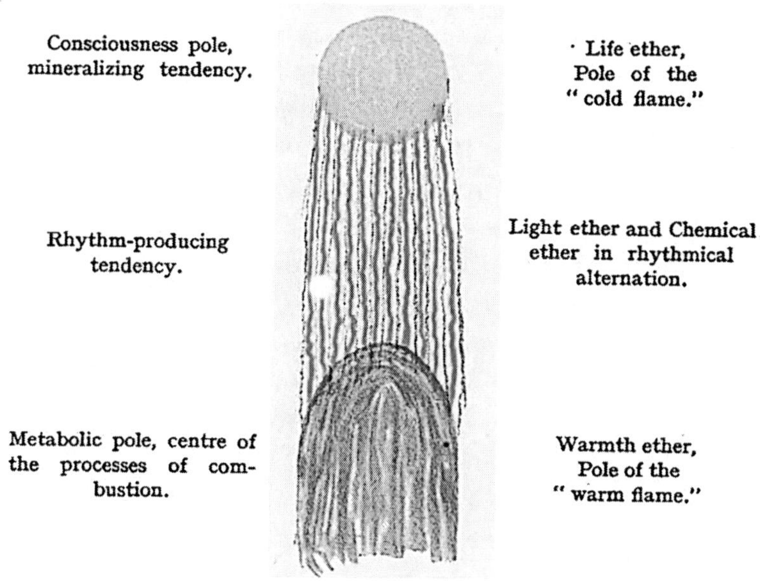

Between the two poles—the upper pole, where the life ether shows its mineralizing tendency in the building of the hard bony structure of the head,—and the lower pole, where the combustion processes of warmth ether have their most important point of focus—there lies in the middle

the rhythmical system of the human organism, as manifested in the breathing and the circulatory process, which has created its central organs in the lungs and the heart, whose rhythms we shall find to be caused by the chemical ether and light ether.

The evolution of the germ begins with a polarity of functions: that is, a separation into two germ-layers, where from the sense instruments are developed out of the outer germ-layer and the metabolic organs out of the inner layer.

When considering the process of heredity, and in explaining the primal predisposition, the "species cell" or "germ cell," that it is the ether body which determines the plan and brings it to accomplishment in the evolving organism. Just as the functions of the germ-layers are polarically systematized in the embryonic process, so also the functions in the growing organism are centralized at two poles: the consciousness pole in the head, which constitutes the center of the nerve-sense processes and also the chief field of action of life ether (cold flame); and on the other hand, the metabolic pole in the lower part of man, the chief field of action of warmth ether (warm flame). If one thus follows the metamorphosis of the evolving organism through the organ-building forces of the ether body consistently from the embryonic state to the completely formed organism, one will understand the living dynamics of the formative forces. One then understands the reality of the phenomena of life.

Already in considering the phenomena of radio-activity, which exert so great an influence upon life phenomena, we have seen that in this way, ozone and hydrogen peroxide are on the one hand formed and on the other hand destroyed. The same thing happens also by action of the blood—that is, the etheric formative forces active in the process of the circulation of the blood. The organism receives the air, the twofold oxygen, brought in through the lungs and the breathing tissues, then changes this into threefold (cold flame, retardation of life process, oxyhemoglobin) and only after a part of the organism has been coursed through in this way does the re-formation take place from 03 to 02 (+ 70 cal. of heat); and hereby comes into manifestation that heat which is characteristic for the heat-processes in organisms (warm flame).

It may be said, therefore, that in the human rhythmic system (breathing, blood circulation, etc.), the light ether works for the upper, consciousness pole; chemical-ether for the lower, metabolic pole, the substance pole; the one formative force works for the physical basis of the spiritual life; the other for the physical basis of the corporeal life. As long as man lives in the substance world, neither of the two conflicting formative forces is victorious, for the living man establishes and maintains an equipoise until the moment of death. *The "life" of man thus manifests itself as the maintenance of an equipoise between conflicting polarities.* Therein lies its nature.

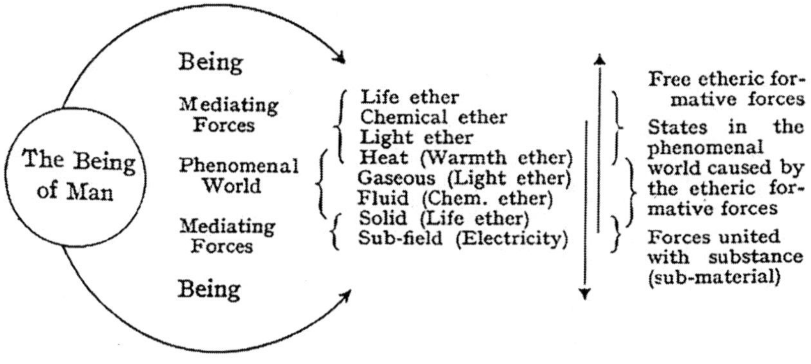

Rudolf Steiner's Ethers Compared to Plasma

"... plasma *is the first state of matter out of which all the other states of matter originated ... Nature began with the* plasma."

Vinod Krishnan

It becomes evident that comparing principles of plasma to Dr. Rudolf Steiner's understanding of the Ethers, and in particular the chemical and life Ethers (the Tree of Life), will show that before plasma cosmologists developed their ideas, Spiritual Science had already defined and predicted the electric nature plasma, this in light of the fact that plasma was first discovered in 1879, the year the Archangel Michael arose to his regency as the current Time Spirit (Archai). Steiner, and his secretary Guenther Wachsmuth, in their book *The Etheric Formative Forces*, clearly spell out the nature of plasma in all its many forms and functions. We have taken some of these Anthroposophic principles on the ethers from that book and will now align them with the information presented in the first chapters of this book. You will notice, they are strikingly similar, leading a clear thinker to realize that plasma is a mechanism of the ethers.

Comparisons

Please note that the top line is a quote from Steiner/ Wachsmuth; the bottom line is from the author Dr. Gabriel:

Ether is simply not matter.

Plasma is not matter.

The cosmos is an ocean of ether.

Plasma is like an ocean that the universe exists in.

The ethers are the bridge between non-spatial and spatial.

Plamsa is not physical matter and fills all of space outside of Earth's atmosphere but also is found on the Earth and in the human body. Thus, it is both non-spatial and spatial.

The spatial comes into existence when the being of a thing passes over into the phenomenal world and mediates the transition from being into phenomenon, and from non-spatial into spatial.

Plasmoids take invisible dust particles and spin them electromagnetically into more complex particles and molecules

The genesis of the etheric formative forces is the genesis of space.

Space, dust, and all space phenomena are plasma. Everything derives from plasma (+ and - primal ionized electrons and protons) except the Earth and its atmosphere and less than 1% of the matter in the universe.

...cosmic ether, fills interplanetary space.

Plasma fills space, creates space, organizes space and supposedly through gravity (life ether) swirls and condensed elemental dust into spheres of plasma called stars, galaxies, etc.

The warmth ether of blood, or blood as the physical bearer of the I/Ego, is dimensionless, non-spatial, and intensive movement.

Plasma is dimensionless, non-spatial and usually in movement. Whether plasma has consciousness is debated and not clear.

Life ether corresponds to the earth element and is the life force of organisms.

Human life has only been found under seven layers of atmospheres that protect biological life from the harmful rays of the sun (ultra-violet, gamma, x-rays, cosmic rays, CMEs, solar flares, and all types of solar forcing; electron, proton, neutron, etc. Cold plasma appears in human blood as the life of the organism.

Life requires chaotic dissolution, upsetting of the general laws tending to fixity.

Plasma defies gravity, and the forces of solid matter, water, thermodynamics, and every other "known force or law" generally accepted by scientists. The pulsing oscillations of plasmoids, as well as their behaviors, demonstrate chaotic internal and external metamorphoses.

Pure light dies under the laws required for heat light. The Sun force forms ozone, threefold oxygen in reciprocal action with the forces and substances of the light ether sphere of the Earth organism. The life ether radiated from the Sun into the Earth organism forms in the light ether sphere ozone, the threefold oxygen.

Direct sunlight is filtered in the ionosphere with the help of the ozone (a plasma) that absorbs the harmful rays and turns cosmic helium into ozone (O3-cold light) which creates a plasmodial ozone layer (atmosphere) that lets light go further down into the troposphere where that life brings warmth and light into the world of humans as part of the basis of life.

The creative forces that are surging inward from cosmic space waken the ever-new spheres of force with exalted rhythms.

Plasma surges into the Earth's magnetosphere and plasmasphere from solar wind (breath), and rhythmic magnetic reconnection between the Sun and Earth every eight minutes. This causes a solar breathing process in the spheres (torus shaped) around the Earth. Certainly, these plasmodial creative forces are "exalted rhythms" since they are at the base of human existence.

The Sun-sphere is the starting point of new force-waves. The ethers are waves of four formative forces surging throughout the cosmos.

The sun is the major source of active plasma for humans. Its rays are both waves and particles. Different types of ether formations throughout the cosmos are explained by the theories of the ethers. Plasma is the scientific explanation for the cosmos. Plasma and the electrical forces of atomic theories.

Steiner formulates thus the fundamental perception: "Will is the 'idea' itself conceived as force."

Some anthroposophists would argue that plasma is the "willpower" of God, that works through the Spirits of the Thrones in the first rank of the divine spiritual hierarchies, the creators of the manifest universe. Some scientists have argued that plasma seems to be the place where "archetypes" are formed and stored.

The important distinction between the oxygen in the threefold state on the one hand and oxygen in the twofold state on the other, will in the future play an important role in any of the phases of the study of organisms.

Plasma is involved in the burst of light at the moment of conception, every heartbeat, everywhere blood goes, and may be part of the mysterious "etheric body" so often spoken of in esoteric traditions. Plasma is in the germ cell, the layers of the skin, every cell, every bio-energetic expression of energy exchange in the body, etc.

Radio-active decay, is the same as that which we discover at the boundaries of our Earth-organism, in the coming into existence of matter: that is, helium (He).

This is the neutral helium (ionized) brought to Earth on the plasma stream from the Sun that charges the plasmasphere with fresh helium elements to spin into molecular ozone (O3) and later into breathable oxygen in the lower atmosphere (O2). Plasma creates what scientists call matter.

In the outer planets, taken collectively, warmth and light ether preponderate. In the inner planets, taken collectively, chemical and life ether preponderate.

The outer planets have radiant warmth and light ether which creates gaseous planets with their matter spun out into rings or moons in the atmosphere around them. No life ether is present to condense matter because outer planets lack the ability to sustain life. The inner planets are more solid due to the presence of condensed plasma or life ether.

There are seven etheric primal forces, formative forces, active in the cosmos; of these, however, only four reveal themselves in the space-and-time processes of our present phenomenal world.

This is why plasma in the human being is somewhat invisible, even though its effects are witnessed and measurable, like so many invisible forces on the Earth. The forces of chemical and life ether may not, at this time, become perceptible to human senses, just like plasma. Plasma is active in atmospheric phenomena regularly but not observed in most cases.

Warmth ether, from which the other ether forces have evolved, has in turn come into being out of purely spiritual states outside of time and space.

Plasma first appeared as dust from some unknown place; or, you can believe in the Big Bang theory, which was a big hot explosion. Space came into being, according to science, from movement—movement is warmth.

The life ether, as a free force, brings about not only the sundering of the Earth organism from the cosmos, its individualization, but also the sundering from the Earth organism of each single body within it.

Plasma steps down cosmic energy into Earth's atmosphere as oxygen and water, which allows humans to come to birth.

Life ether is the force of gravity. It takes over as the Earth force; it is individualized. Life ether is the uniform common cause of gravitation and terrestrial magnetism.

Plasma (life ether) creates celestial bodies, including Earth. Life ether works through plasma forces (organizes electricity). Science says the universe is almost entirely plasma and its electric forces and the force of gravity.

Warmth ether (heat) and Light ether (gaseous substances) have anti-individual form tendencies, Chemical ether (fluid substances) have partly anti-individual, partly individualizing form tendencies, whereas life ether (solid substances) have purely individualizing form tendencies.

Plasma descends into matter in different forms.

Electric and etheric currents encircle the globe.

Plasma follows electrical currents like the Van Allen belts.

Considered from the etheric point of view, the Sun is a space not filled with substance, but in which the etheric forces of life ether are active.

Scientists have now theorized that the Sun is a white-hole of plasma.

The gaseous state—that is, the surrounding air, the atmosphere where light ether strives for a rarefaction of substance; the chemical ether for a condensation.

Plasma creates and is very active in atmospheric phenomena.

The evolution of the world takes its course, therefore, not merely between primal nebula and heat-death, but between space-genesis and space annihilation—between the affirmation and the denial of space.

New theories of the universe place plasma as the creator, not the primal nebula. Plasma is not winding down, it is ever creative.

It is by no means only "electrons" that stream in eternal alternation into the human organism and the living beings of the other kingdoms of nature and out again, but that these are entities that, in their totality, can never be grasped within the range of the theories of modern physics.

Plasmas are often electron based and do go in and out of humans in many ways.

The inward drawing forces of the chemical and life ethers, tending to concentrate, have worked in toward the center of the earth, produced the solid and fluid substances, and assembled these toward the middle spheres of the earth organism but also in the formation and union of substance.

Plasmas are theorized and observed to create oxygen, water, and solids in complex dusty plasmoids,

The solid Earth with its atmosphere—is formed so that it consists of four globular concentric spheres, and in each of these four spheres, there is one of the four etheric formative forces.

Plasma helps create and sustain the magnetosphere, plasmasphere, ionosphere, ozone sphere, and the combined layers of the atmosphere.

The outward-tending, centrifugal forces of light ether and warmth ether have their spheres of activity primarily in the gaseous atmosphere surrounding the globe. This gaseous atmosphere is surrounded by a mantle of warmth ether at the outer borders of the atmosphere, towards the cosmos.

Plasmas are extremely active in the super-hot thermosphere.

It is that which is rayed out to us, among other things, from the Sun, and then modified in its action by the atmosphere of the Earth. Life ether, together with chemical ether, belongs to the group of suctional forces: those which tend to draw inwards. It is related to that which is called "gravitation" and to the phenomenon of magnetism.

Plasmas can generate electricity, magnetism, and defy gravity.

The magnetic poles are points of the least disturbance, and hence of greater concentration and activity in the force-sphere that belongs to the force of life ether.

Plasma is most active in our atmosphere at the magnetic poles.

Magnetism and gravitation, both on the Earth and in the cosmos, are forms of expression of the same cause—namely, of the life ether. The Sun is a cavity not filled with substance but is a "minus space" or a "vacuity in space," as contrasted with the rest of world space, with its substantial and insubstantial contents.

The sun is an archetypal plasma ball, a plasmoid, and it is "empty" inside and yet can create temperatures of over a million degrees right next to segregated areas of super-cold. Fusion reactions take place in these sun-like balls of fire—the genesis of matter.

The air in the spheres of the Earth organism are filled with innumerable tiny vacuums (after the manner of the Sun) in which there are suctional forces that, when they issue from these vacuums, cause a drawing together or a condensing of the surrounding substance (water vapor). A condensing process can be established with rhythmic repetitions through which the etheric formative forces cause the genesis of substance, its maintenance, and its reduction—that is, its metamorphosis into another state.

Here is the earliest theory of the genesis and continuous creation of water in our atmospheres in tiny vacuums (plasmoids).

Man, by means of his etheric forces, carries out within himself, in his blood currents and in his breathing, the processes of combustion and of that peculiarly labile transformation of oxygen. Something of the nature of the flame coming into existence under certain conditions and then dying out is also active in man. The human being bears within himself the secret being of the "flame" of fire, of which the myth of Prometheus speaks. Pure light shines as a cold flame. Its action is free from heat.

An excellent description of cold plasma that can create a flame. This is the activity of plasma in blood.

A gaseous substance always tends to break through the form imposed upon it; it tends to release itself into the world in every direction. Gas opposes the earth's laws of form. It will not submit, like the other states of matter, to the earth relationships and tends away from the laws of Earth in order to return to cosmic laws. It denies the sundering, individualizing process toward which Earth tends, as opposed to the cosmos: its tendency is anti-individual.

Without magnetic force lines to hold plasma in lower atmospheres, it would rise back into space like a gas since it often acts like a gas.

If it is principally the life ether working upon a certain part of the organism (the lymph system, the center of gravity, the life ether sphere), then the so-called colorless (white) blood corpuscles appear.

Plasma in the atmosphere acts like white blood cells in the human body, protecting against outside forces that harm the body.

The blood causes a transitional metamorphosis from the laws of the "warm flame" to those of the "cold flame"; it thus checks the rapid combustion within the organism and renders the slower process of combustion in the living being possible.

Plasma acts as a step-down transformation of hot material into cold substance, especially water.

Radioactivity exerts a great influence upon life phenomena as ozone and hydrogen peroxide are formed and destroyed. This also happens because of the etheric formative forces active in the process of the circulation of the blood. The organism receives the air—the twofold oxygen—through the lungs and the breathing tissues. It changes this into threefold oxygen ("cold flame," retardation of life process, oxyhemoglobin), and only after a part of the organism has been coursed through in this way does the re-formation take place from O3 to O2 and hereby become heat which, is characteristic of the heat-processes in organisms.

Ozone and hydrogen peroxide is found in white blood cells in the body, carried by blood plasma and delivered to any pathogen found in the blood.

We can study the configuration of the etheric body of the Earth, which is a reflection of our own inner being, and we must realize that the etheric forces molding the body of the Earth are the same as those in our blood corpuscles. When a change takes place in the etheric structure of the human blood, this will be reflected in the etheric sphere of Earth, because Earth and the blood corpuscles are formed according to the same laws.

Plasma generated energy in the heart enters the blood stream and creates a magnetic field around the human body. This plasmodically generated energy is in harmonic resonance with the magnetic fields of the Earth and is the signature of life of Earth—the monad of individuality.

We shall never be able to reach a true conception of ether by transferring the forms and laws of motion in matter to the ether itself. If, however, we conceive of ether, or the etheric formative forces, as formative forces void of any quality of substance, as active principles which come to living expression in the phenomenal world only through their active tendencies to definite motions, to shaping definite forms with definite qualities, then this difficulty disappears.

Plasma shapes and creates form in matter but is not matter.

The etheric primal forces, formative forces, do not belong immediately to the world of phenomena, and they are, like all forces, imperceptible to the physical senses; they belong to a supersensible set of facts. They are to be understood only when we consciously bear in mind that something real, the very beings of things, comes with these forces to living expression in the phenomenal world.

Plasma is often invisible and some theorists believe it is a living Being.

Life ether is rayed out to us, among other things, from the Sun and then modified in its action by the atmosphere of the Earth.

The same is true for plasma.

We can never understand the phenomena of cosmos and Earth, of atmosphere, atmospheric electricity, unless we view the Earth as a great living organism which, like all living organisms, is characterized by a process of breathing, circulation, etc.

> Plasma comes in through the breath of the solar wind and feeds a living system.

Naturally, life ether in small quantities enters also into the atmosphere; indeed, it is this which holds the atmosphere together by its suctional action, since otherwise—in accordance with the tendency of light and warmth ether and the gaseous substances adapted to these—the atmosphere would fly away into cosmic space.

> Plasma is bound to the Earth by the Van Allen magnetic field lines.

The difference between the Sun and the inner Earth consists in this, that in the Sun the life ether works only as a free etheric force and therefore induces the tremendous magnetic intensity of the Sun, whereas in the inner Earth it works reciprocally with substance and therewith produces the solid state of aggregation of the fourth sphere (shell of the Earth) adapted to the life ether, or else produces modifications of this through the co-operation of the other formative forces toward the center.

> This is the difference between plasma working in matter or in matter-free space.

A condensing process—with rhythmic repetitions—can be established, through which the etheric formative forces cause the genesis of substance, its maintenance, and its reduction, that is, its metamorphosis into another state.

Plasma, just as the ethers, creates matter and can metamorphose it into other substances.

The phenomenal world of ponderable substance is a medium state extending between the force activities of the etheric on one side and electricity on the other. The state of substance at any time is the result of this conflict between forces over the mutable world of substance.

Electric plasma on one side, and the forces of the ethers on the other, with ponderable substance in the middle.

The etheric formative forces are the bridge between being and phenomena, but thereby at the same time the bridge also between non-spatial and spatial. For only when they begin to work does there originate a juxtaposition and a succession that is, space and time.

Plasma is non-spatial and perhaps a Being.

The man of today, then, if he were living in the upper light ether sphere of the Earth organism, would be an entirely differently constituted being, with quite different organs of perception and a correspondingly different state of consciousness.

Some would suggest that plasma is a higher state of consciousness.

In man also there is active something of the nature of the flame coming into existence under certain conditions and then dying out, something of the polarity of "heat-light" and "pure-light." Man bears within himself the secret being of the "flame" of fire, of which the myth of Prometheus speaks.

Plasma brought electric fire down to the Earth.

Like the cosmic formative forces which penetrate into the outer spheres of the Earth organism, life ether not only forms the ozone, so vitally important, but reaches deep into the phenomena of life. The physiological effects of these released forces are almost all-comprehensive.

This is what they say about the future of plasma technologies—all comprehensive. Did plasma consciously create the ozone layer to protect its offspring? Is plasma a god?

The model of the Earth organism is also the model of the smallest units of substance atoms, which, however, go through their metamorphoses according to the different influences of the various regions of the Earth and thus are more and more varied and individualized in space and time.

Plasma may create what science calls atoms.

The predisposition of the species cell—that is, that which is inherited in the embryo, which determines by law what it will become—is the ether body. The essential, therefore, in heredity is not the substance of the germ cell but its etheric structure—that is, those etheric formative forces which chiefly determine the coming into existence of this organism.

Is the light created at conception
plasma light—the ether body?

What has brought to pass the genesis of the macro-organism, this brings about also the genesis of the embryo.

Primal plasma in space (cosmogenesis) and
plasma in the embryo (anthropogenesis).

The evolution of the germ begins with a polarity of functions: that is, a separation into two germ-layers, where from the sense instruments are developed out of the outer germ-layer and the metabolic organs out of the inner layer.

Plasma balls, cells, and plasmoids
have double walled layers.

When considering the process of heredity, and in explaining the primal predisposition, the "species cell" or "germ cell," that it is the ether body which determines the plan and brings it to accomplishment in the evolving organism. Just as the functions of the germ-layers are polarically systematized in the embryonic process, so also the functions in the growing organism are centralized at two poles: the consciousness pole in the head, which constitutes the center of the nerve-sense processes and also the chief field of action of life ether (cold flame); and on the other hand, the metabolic pole in the lower part of man, the chief field of action of warmth ether (warm flame).

Is a human germ cell a plasmoid
with an archetypal DNA sequence?

Ozone and hydrogen peroxide are on the one hand formed and on the other hand destroyed. The same thing happens also by action of the blood—that is, the etheric formative forces active in the process of the circulation of the blood. The organism receives the air, the twofold oxygen, brought in through the lungs and the breathing tissues, then changes this into threefold (cold flame, retardation of life process, oxyhemoglobin) and only after a part of the organism has been coursed through in this way does the re-formation take place from 03 to 02 (+ 70 cal. of heat); and hereby comes into manifestation that heat which is characteristic for the heat-processes in organisms (warm flame).

Plasma can form as extremely hot or
extremely cold. It is cold or hot flame.

The "life" of man thus manifests itself as the maintenance of an equipoise between conflicting polarities.

Between the polarities of electric plasma
and the etheric formative forces.

Hydrogen Peroxide makes Water and Ozone makes Oxygen

In Earth's atmosphere, hydrogen peroxide plays a key role in the primal chemistry of water, just as ozone does for oxygen...

In the human body, hydrogen peroxide is primarily found in the lungs, gut (especially the oral cavity and stomach), and thyroid gland; however, it can be present in most cells due to its production within mitochondria, with some tissues like the vascular endothelium and circulating blood cells also experiencing higher concentrations of hydrogen peroxide. Most cells produce hydrogen peroxide as a byproduct of cellular metabolism, particularly within the mitochondria. The immune system utilizes hydrogen peroxide as a signaling molecule to fight infection, particularly in the lungs and gut where it helps activate white blood cells. While present throughout the body, certain tissues like the oral cavity, esophagus, stomach lining, and respiratory tract may have higher levels of hydrogen peroxide due to their exposure to oxygen and potential microbial activity.

The white blood cell that contains hydrogen peroxide is a neutrophil. Neutrophils are a type of phagocytic white blood cell that actively fights bacteria and fungi by releasing chemicals like hydrogen peroxide from their granules. They are the most abundant type of white blood cell and are considered the "first responders" to infection. Neutrophils engulf pathogens and then release enzymes and reactive oxygen species, including hydrogen peroxide, to destroy them.

Hydrogen peroxide (H2O2) is highly toxic inside most cells and is used by our immune system to kill infecting bacteria. Hydrogen peroxide produces highly reactive hydroxyl radicals. When ozone and hydrogen peroxide react, ozone oxidizes hydrogen peroxide to water, and in the process, ozone is reduced to oxygen gas. Ozone reacts with H2O2 to produce oxygen. One volume of ozone gives a half volume of oxygen. Catalase is an antioxidant enzyme found in all aerobic organisms that breaks down hydrogen peroxide into water and oxygen.

Hydrogen Peroxide Found in Space

Published in the journal *Astronomy & Astrophysics*, by ESO, July 6, 2011

The colorful Rho Ophiuchi star formation region, about 400 light-years from Earth, contains very cold (around -250 degrees Celsius), dense clouds of cosmic gas and dust, in which new stars are being born. The clouds are mostly made of hydrogen, but contain traces of other chemicals, and are prime targets for astronomers hunting for molecules in space. Astronomers using the APEX telescope to observe this region discovered hydrogen peroxide molecules in interstellar space for the first time. The discovery gives clues about the chemical link between two molecules critical for life: water and oxygen. *On Earth, hydrogen peroxide plays a key role in the chemistry of water and ozone in our planet's atmosphere* and is familiar for its use as a disinfectant. Now it has been detected in space by astronomers using the ESO-operated APEX telescope in Chile.

Hydrogen peroxide (H2O2) is a key molecule for both astronomers and chemists. Its formation is closely linked to two other familiar molecules, oxygen and water, which are critical for life. Because much of the water on our planet is thought to have originally formed in space, scientists are keen to understand how it is created.

Hydrogen peroxide is thought to form in space on the surfaces of cosmic dust grains — very fine particles similar to sand and soot — when hydrogen (H) is added to oxygen molecules (O2). A further reaction of the hydrogen peroxide with more hydrogen is one way to produce water (H2O). This new detection of hydrogen peroxide will therefore help astronomers better understand the formation of water in the universe.

"We don't understand yet how some of the most important molecules here on Earth are made in space. But our discovery of hydrogen peroxide with APEX seems to be showing us that cosmic dust is the missing ingredient in the process," says Bérengère Parise, head of the Emmy Noether research group on star formation and astrochemistry at the Max-Planck Institute for Radio Astronomy in Germany, and a co-author of the paper.

Ozone Therapy

Ozone is a bio-oxidative treatment that promotes healing by oxygenating the organism. Ozone (O3) is a colorless gas composed of three oxygen atoms and acts in the body through oxidation and oxygenation. When ozone is introduced into the body, the third oxygen atom binds itself to harmful microorganisms. By doing so, it oxidizes or destroys them. This process is called oxidation. Ozone then converts to oxygen (O_2) in the body, increasing blood oxygen levels. This process is called oxygenation. Ozone is applied to kill microorganisms, improve cellular function and promote healing of damaged tissues. Ozone therapy is low-risk and is a complementary, supportive and reconstructive form of treatment, often accompanied by standard medical treatments. Ozone therapy can be administered in several ways, including:

Autohemotherapy (ozonizing the blood): The contact of the patient's blood with ozone gas in autohemotherapy allows the red blood cells to transport oxygen more efficiently, which increases cellular respiration. The body receives more energy and the body's natural immunity improves. The number of cytokines that help inflammation decreases and in turn the number of interferons and other "healing" cytokines increases. Autohemotherapy is administered in two ways. In major autohemotherapy, about 400 ml of the patient's own blood is drawn, then enriched with ozone and reintroduced into the patient's vein. Ozone reacts with the blood and forms different types of compounds that produce immediate therapeutic effects. Therefore, with 500 ml of oxygen/ozone mixture, ozone can react with almost all the blood in the body. In minor autohemotherapy, about 5-10 ml of the patient's own blood is drawn, then enriched with ozone and injected into the patient's muscle, usually the gluteal muscle.

Ozone injections: It is possible to inject ozone gas into muscles and joints. In these cases, the amount of ozone is very small (5-10ml). This method is usually used for local symptoms. The most commonly used are intra-articular injections. It is also used for trigger point injections into muscles.

Insufflation: This method involves blowing ozone into a body cavity such as the mouth, ear, rectum or vagina using a special device.

Ear insufflation: Ozone is infused through a catheter inserted into the ear canal so that the ozone gas enters the lymphatic and blood system. Many patients have found success in alleviating allergies, relieving colds, sore throats and swollen glands.

Ozone steam sauna: It is a combination of ozone therapy and steam sauna therapy. Saunas are one of the most effective detoxification techniques as they trigger sweating, which increases blood circulation, reduces inflammation and pain and removes toxins through the skin. When we combine a steam sauna with ozone therapy, it allows ozone to enter your skin through the bloodstream. The moist air opens your pores and allows ozone to enter your body more easily. Ozone enters the bloodstream, fat and lymphatic system through your skin.

Prolotherapy: It consists of ozone injections to the same area following the prolotherapy procedure.

Wound bagging: The wound is sealed with an ozone-resistant bag and exposed to the gas. It is used in non-healing wounds and diabetic feet, skin lesions, infections, circulatory disorders, neuropathic pain.

Hydrogen Peroxide Therapy

Hydrogen peroxide IV therapy involves introducing a diluted solution of hydrogen peroxide directly into the bloodstream through an IV. The goal of IV hydrogen peroxide therapy is to increase oxygen levels in the body, potentially aiding in detoxification, boosting the immune system, and improving overall health. The mechanism of action involves hydrogen peroxide breaking down into water and oxygen once inside the body. This process releases oxygen at a cellular level, which is thought to help improve oxygenation and support cellular respiration.

Enhanced oxygen levels might help in fighting infections, reducing inflammation, and boosting a person's overall health. The increased oxygen may also help detoxify the body by breaking down toxins and waste products, making IV hydrogen peroxide therapy a potential option for anyone who is interested in exploring alternative treatments.

One of the main hydrogen peroxide IV therapy benefits is its potential to detoxify the body. When hydrogen peroxide breaks down into water and oxygen in the bloodstream, the extra oxygen helps to break down toxins and waste products. This may make it easier for the body to eliminate these harmful substances. As a result, some people feel more energized and clearer-headed after hydrogen peroxide IV therapy. The detoxification process may improve overall health and vitality, making it a popular choice for anyone who is looking to cleanse their body and boost their well-being.

Are Plasmoids Biological Life?

"And I looked, and, behold, a whirlwind came out of the north, a great cloud, and a fire enfolding itself, and a brightness was about it, and out of the midst thereof as the color of amber, out of the midst of the fire."

Ezekiel 1: 4

The research presented below is a convincing argument that complex dusty plasmoids may be organic life instead of simple reactive electromagnetic phenomena. We have described above the many known characteristics of plasma that science finds in space, Earth, and Earth's atmospheres and can see that science, in general, does not believe plasma, in any form, is alive. The research study below disagrees because all of the general characteristics that science attributes to living organisms appear in complex dusty plasmoids. We will leave it up to the reader to determine for themselves whether plasma is alive or just seems to be alive.

From: *Journal of Modern Physics*, January 2025, Volume 16, *Plasmas: 4th Domain of Life Plasmas: A Fourth Domain of Life? RNA, DNA and Statistical Analysis of "Unidentified Anomalous Phenomenon" in the Thermosphere,* Rhawn Joseph, Richard A. Armstrong, Konrad Wolowski, Mustafa Abu Safa, Rosanna R. del Gaudio, Rudolph Schild

https://www.researchgate.net/publication/385596004_Plasmas_A_Fourth_Domain_of_Life_RNA_DNA_and_Statistical_Analysis_of_Unidentified_Anomalous_Phenomenon_in_the_Thermosphere

Abstract

The data presented here challenge all conceptions of what constitutes "life" and the origins of life. Self-illuminated plasmas (AKA plasmoids/UAP), with a nucleus and double cellular layers, engage in complex behaviors in the thermosphere and display what appears to be mutual awareness, purposeful contact-seeking and mitosis and ejection-secretion of interactive plasmoid-clouds of particles and additional plasmoids that seek out other plasmas. Plasmas also engage in communicative signaling via oscillations in size and illumination, with some oscillating relatively rapidly as they approach and others relative slowly after they pass by other plasmas. Plasmoids turn, follow, target and collide with and/or merge or pierce other plasmas, whereas yet others form thick glowing plasma bridges linking four or more plasmoids together, reminiscent of interactions between colliding galaxies and colonies of algae. Colliding plasmoids traveling at high velocities also release glowing plasmatic clouds in their wake, reminiscent of a comet's tail. That pattern repeat is predictable given that plasmas represent a fourth state of matter, and that over 95% of the universe consists of plasma in various states. As previously reported: Plasmoids that descend into the lower atmosphere are likely to be perceived as UAP/UFOs which engage in behaviors similar to plasmas in the thermosphere. A quantitative analysis determined unique behavioral characteristics of plasmoids that were filmed by NASA space shuttle mission STS 75 swarming toward and gathering around a Tethered Satellite System (TSS-IR) positioned 296 Km above the Earth. Analyses were made of 91 consecutive freeze-frames taken every 0.2s. It was determined that plasmas accelerate to exceptional velocities (up to 35.6 Km s-1) and make abrupt 163° turns in trajectory. Four 'collision events' were analyzed in detail and all plasmas altered velocity, trajectory, and/or shape either before, during, or after the collision event. Additional analysis revealed networks and chains of filamentary plasmatic magnetic flux ropes and cables in the thermosphere and resembling vast neural networks and which may be producing plasmoid entities whereas ganglia-neural networks have also been detected in some plasmoid specimens. Dusty plasmas may have provided the internal environment for the genesis of RNA, then DNA and the origins of life via the assimilation of all necessary elements which are available in space; and may have fashioned a plasma genome via the incorporation of the genomes of bacteria, algae, fungi, lichens, etc. propelled into the upper atmosphere by bolides, hurricanes and powerful winds. Hence, plasmas engage in complex behavior possibly because they have a genome and are alive and sentient and represent a fourth domain of life. However, what appears to be life-like behaviors may be due to differential electromagnetic activity, density, and negative vs positive electrical charges and push-pull magnetic repulsion and attraction as similar interactions are observed between galaxies, pulsars, binary stars, and red giants.

Introduction

Glowing, pulsating, self-illuminated plasmas in the thermosphere engage in complex behaviors and interactions that appear life-life and purposeful. These interactions include what appears to be communicative signaling via oscillations in size and illumination, with some pulsating relatively rapidly as they approach, and others relative slowly after they pass other plasmas. These plasmas are multi-layered electromagnetic cellular entities, up to several Km in size and have been repeatedly observed by NASA space shuttle crews and captured on film by ten different space shuttle missions engaging in gravity-defying maneuvers and complex behaviors, and approaching and lingering near the MIR International Space Station and a number of space shuttles including appearing outside windows and upsetting the crews. Plasmoids that descend into the lower atmosphere are likely perceived as UAP/UFOs.

As detailed in this report, dusty plasmas may have provided the internal environment for the genesis of RNA then DNA and the origins of life via the assimilation of all the necessary elements which are available in space; and may have fashioned a plasma genome via the incorporation of the genomes of bacteria, algae, fungi, lichens, etc. propelled into the upper atmosphere by bolides, hurricanes, and powerful winds. We also hypothesize that plasmas engage in complex behavior because they are alive and sentient and represent a fourth domain of life. A quantitative analyses of plasmoid behavior, size, movement, velocity, and contact seeking supports that hypothesis. We also report the discovery of extensive networks of plasmodic magnetic flux tubes, ropes and cables in the thermosphere that may be giving birth to or providing energy to plasmoids.

Some plasmas appear to behave as if they are swimming in water, or rather, in a sea of electricity. It has also been noted that plasmas in the thermosphere, and their behaviors and multi-plasma cellular composition, bear similarities to water dwelling algae, diatoms and dinoflagellates

Based on analysis of shape, size, and complex behaviors, it has been proposed that the plasmas of the thermosphere could be classified as "electromagnetic extremophiles" and "Hunters" "Grazers" and "Floaters" whereas "Sinkers" could be likened to "Thunderstorm Divers" that descend into thunderstorms and the lower atmosphere where they are likely to be classified as UAP/UFOs.

It has been documented that these self-illuminated plasmas of the thermosphere have a variety of shapes, e.g. cloud, cone, ring, donut, serpentine, oscillate in size and illumination, will suddenly materialize and grow in size, speed up and slow down and commonly make gradual or sudden and rapid 45, 90, 180 degree turns and will follow or collide or pierce other plasmas; the collisions often associated with the release of Km in length cometary-like plasma trails just prior to or following collisions. Some plasmas in the thermosphere also consist of conglomerates of multiple plasmas; and in actions reminiscent of cellular mitosis, some will divide and replicate or secrete-eject plasmoids that follow, confront, contact and interact with other plasmas. In addition, some plasmas merge with and/or form networks of plasma links with other plasmas.

Plasmas in the thermosphere often target and collide sometimes head-on with another plasmoid such that two rapidly moving plasmoids appear to purposefully seek out each other. Not uncommonly, one or both will eject, in their wake, a cloudy illuminated stream of plasma particles and dust or smaller plasmoids prior to or following the collision. In some instances, a plasmoid that is being followed will also eject, in their wake, a smaller plasmoid that will target the plasma that is following. This ejected mini-plasmoid will strike and/or just prior to collision, cause the plasmoid that is following to suddenly reverse direction! Hunters will also sometimes eject a plasmoid replicon in their wake, and sometimes the replicon will remain relatively stationary (like a Floater) as the Hunter continues to hunt. Some plasmoids will also merge for several seconds before moving on. Not uncommonly, a cloud of plasma will briefly connect two merging or mutually contacting plasma, just before they make contact and as they disconnect, perhaps reflections of electromagnetic interactions. Yet others will form multi-plasma bridges that link all together. Whether these are purely electrical exchanges or evidence of purposeful social behavior, is unknown. Plasma will also change from a dark mode to illuminate, or from illuminated to dark mode. Yet others will suddenly materialize and grow in size as they change shape. Commonly, plasmas change shape and size as they move through space or approach one another.

Investigators report that plasma created in a laboratory engage in life-like purposeful behaviors and complex interactions, including tracking, merging, and colliding (sometimes referred to as "collisionality" and "energy cannibalism"). Identical behaviors have been observed in the thermosphere. Navy pilots, who observed and filmed these plasma-like forms, claimed these entities display intelligence and an awareness of their surroundings. Plasmas in the thermosphere approach and congregate near sources of electromagnetic activity including an electrified tether generating electromagnetic pulses into the space medium and above thunderstorms where many will approach, pierce, merge and pass by each other. They also approach these storms at varying speeds only to slow down as they descend into the storm without burning up. Yet others will ascend from thunderstorms and hurricanes back into the thermosphere and may pace or follow or approach the space shuttles.

These plasmas are electromagnetic entities and are obviously attracted to sources of electrical activity. It has been hypothesized that plasmoids which have descended into the lower atmosphere account for many observations of UAP/UFOs, especially those seen above nuclear power plants and nuclear-powered ships and sites of nuclear disasters and explosions; a thesis that challenges all traditional extraterrestrial explanations; provides a scientific framework for understanding some phenomena associated with UAP and which effectively ties together atmospheric science (e.g. thunderstorms, lightning, atmospheric charging and white water ocean conditions) with observations of UAP from Navy pilots and as based on U.S. Navy film footage. It is now well documented that some UAPs exhibit complex behaviors akin to those typical of plasmoids in the thermosphere and may be responsible for what is referred to as "inexplicable" airline disasters. The fact is: These entities (UAP/UFO) once they descend into the lower atmosphere and like those in the thermosphere, display no means of propulsion, make "impossible turns," approach and nearly collide with aircraft, and can suddenly accelerate to gravity defying hyper-velocities and disappear into the upper atmosphere. This is typical plasma behavior in the thermosphere, and characteristic of plasmas (UAP/UFOs) in the lower atmosphere as documented by Navy pilots Ryan Graves and David Fravor in prepared statements for the United States Congressional Oversight Committee.

And just as plasmoids in the thermosphere commonly follow, target and collide with each other, Navy pilots have reported numerous encounters where they had to take evasive action to avoid collisions with UAP which more often than not, have an oval form and change shape and possibly internal layers similar to plasmoids.

In an earlier report, it has been documented that increased sightings of UAP also correspond with increased incidents of inexplicable aircraft disasters—a function possibly due to plasma induced electronics failure and mental disturbances, and/or head-on collisions. It is well documented that encounters with these plasmoid-like UAPs can short-out electronics, radar, and auto-pilots. The disruption of electronics is predictable given that plasmoids are electromagnetic entities that are attracted to and emit electromagnetic radiation.

It can be predicted that plasmoids (UAP/UFOs) observed by Navy pilots defy "material science" because plasmas obey the laws of electrodynamics and are not bound by technology or gravity. Moreover, plasmas are excellent electrical conductors, similar to wires carrying current. These currents produce their own magnetic fields which generate currents with velocities that can range from 5 to 50 km/sec. Plasmas obey the laws of electromagnetism and commonly accelerate to hyper velocities and engage in gravity-defying behaviors. Not surprisingly, plasmas that descend into the lower atmosphere can easily outmaneuver military aircraft and fly off at incredible velocities.

Plasmas in the thermosphere are distinct from those created in a laboratory as they are kilometers in size and endure for long, albeit unknown lengths of time. These double, often multi-layered dusty plasmoids of the thermosphere have likely incorporated gases, dust, grains, and fragments of meteorites, as well as biological matter cast into upper atmosphere by powerful winds and storms—and all of which may be charged to varying degrees, thereby forming interconnected circuits that carry a charge, as well as producing electric currents that flow into the double layers. As proposed by Alfven, kinetic energy (of the convection) is converted into electromagnetic energy that flows throughout the circuit and generates electrostatic energy in the double layers which emit particle beams of energy and "high energy electrons" which radiates light. Therefore, plasmas/ plasmoids in the thermosphere are self-illuminating because dust, particles, and electrons within or beneath their double cellular layers are highly charged. When these particles and dust become highly excited, they collide and the energy released includes photons thereby illuminating the interior and surroundings of the plasma.

For example, electric current from the sun flows into the upper atmosphere and excites plasma to such a degree that it will glow. The magnetospheres of planets have depressions, holes above their magnetic poles, allowing electric current from the sun to gain entry to the upper atmosphere and triggering brightly colored plasmas, e.g. auroras. Auroras (also known as the northern lights and southern lights) have a variety of colors including blue, red, yellow, green, and orange which flow and shift gently and change shape as they flow. Auroras are plasmas. Colors depend on what internal gases are being ionized, which is why auroras have a variety of hues. Plasmas created in the laboratory with high intensity electric currents also change shape and will glow (as do those in the thermosphere), the brightness depending on the intensity of the strength of the current. Plasmas in the thermosphere not only glow with light, but many will oscillate with light in repetitive heart-beat rhythms of brightening then dimming. Plasma circuits naturally oscillate. However, different plasmas in the thermosphere often display different patterns and rhythms, whereas not all plasmas will oscillate. Moreover, some, but not all plasmas/ plasmoids oscillate with light as they and other plasmoids approach thunderstorms or each other

Plasmoids appear to communicate via oscillations in size and illumination, with some oscillating between bright and dim and from a larger to a smaller size relatively rapidly as they approach and others relatively slowly after they pass by other plasmas. Yet others that were illuminated seemingly disappear, presumably because they "turned off" their light or flew off into deep space, whereas others may suddenly or rapidly materialize, possibly because they were in a dark mode and turned on their lights, or because they erupted from a hole in space-time. Those that are "cone-shaped" are the most likely to alter their patterns of size and illumination; behaviors that resemble rhythmic pumping.

Not all plasmas will emit light. The ionosphere is a plasma that does not emit light except during auroras. Much of the universe consists of plasma, which is in a dark mode. Furthermore, shadowy dark shape-shifting forms have been observed among plasmoids congregating near the electrified tether filmed by STS 75. Some of these latter plasmoids have also been noted to glow as they approach and contact other plasmas and this may be due to increases in electric current and voltage across their layers. Or, it may serve a communicative function and is purposeful.

Plasmas operate in a dark mode, normal glow mode, arc mode and oscillate with light. The brightness of the glow likely depends on current intensity and plasma density. The stronger the current, the brighter the plasma. If current density increases and a higher voltage passes across plasma it will light up. However, if the voltage drops, the plasma will also light up. This is due, in large part to the movement of electrons. Electrons, being less massive than ions, also move about freely, back and forth around the more slowly moving ions, creating harmonic oscillations and produce light when they move between energy levels within an atom. When an electron "jumps down" to a lower energy level, it emits a photon of light with a specific wavelength and color depending on the energy difference between the levels. Different energy level transitions produce different wavelengths of light, which can be perceived as oscillations as the light dims and brightens. Hence, plasmas respond to continued increases in current with a rise then a drop then a rise in voltage and the plasma will oscillate in size and with light. As current continues to increase, the plasma will jump into an arc mode and shine with high intensity bright white light.

It could be argued, therefore, that plasmas are responding to the charges and electrical circuits in their immediate environment. That environment includes plasmas with different densities and electric charges. However, this explanation cannot account for why plasmas also are also expanding and shrinking in size as they move through space; and why those adjacent one another show different patterns of oscillation; and why those approaching vs passing by other plasmas oscillate at different speeds. These latter observations support the hypothesis that some plasmoids may be communicating via a "language of light" i.e. oscillating electromagnetic waves and that the shape-shifting pumping patterns of shrinking and expanding are a means of purposeful locomotion. Given that plasmas also display a variety of shapes and behaviors, and the fact that not all oscillate in size, shape, and illumination, also further documents that a variety of plasmoids dwell in the upper atmosphere and that diverse species of "life" swim the seas of space surrounding our planet.

Langmuir was the first to use the word "plasma" to describe those created in a laboratory. He observed that the fluid-like behavior through which high velocity electrons and ions and plasmas were moving was similar to the way blood plasma carried white and red corpuscles. He also noted that when a charged object is introduced, the plasma will form cellular structures and double layers. Hence, the term "Langmuir sheaths" refers to the double layers and cell walls typically formed by plasmas. That plasmas in the thermosphere have double (and often multiple) layers, voids, and what resembles a nucleus, has also been documented as also reported here. It is unknown if these layers and cell walls consist solely of plasma or a plasmoidic concentration of substances commonly found in space and in the upper atmosphere such as oxygen, carbon, and hydrogen and the remnants of carbonaceous chondrites and organic-biological matter.

Lipid bilayers are permeable to amino acids and would enable dusty plasmas to incorporate the shattered remnants of carbonaceous chondrites and all the nucleotides, acids, proteins necessary for life and the fashioning of RNA and DNA. However, plasma, in itself is permeable, which would allow all these chemicals and molecules to become incorporated. As will be detailed, all the ingredients and conditions necessary for building complex molecular organic structures, amino acids and proteins are present in space and in nebular clouds including phosphorus, calcium, water, carbon, and oxygen; which when mixed together and when irradiated might easily produce self-replicating carbon-crystal-helixes. In fact, all the necessary elements, chemical, amino acids and nucleotides necessary for the establishment of RNA-DNA, and thus life, are abundant in the vast environment of space and the upper atmosphere; and which would likely be incorporated within dusty plasmas via their semi-permeable plasma membranes. Hydrogen, oxygen, carbon, calcium, sulfur, nitrogen and phosphorus incorporated within a dusty plasmas would be continually irradiated by ions thereby generating small organic molecules which would then evolve into larger complex organic molecules thereby resulting in the formation of amino acids and other compounds. Moreover, the presence of hydrocarbons (catalyzed by stellar radiation), once incorporated, could be employed for polymerization and the additional creation and assembly of the necessary elements and macromolecules essential for life. Further, in the presence of multiple sources and types of energy, radiation, and hydrogen gasses, these double layered membranes would likely evolve capacitance with hydrogen, oxygen or other gasses acting as an additional energy source. Given that energy and all the necessary life-sustaining molecules, acids and nucleotides may be commonly incorporated within a dusty plasma, it is not unreasonable to assume that this mix would eventually lead to the creation of at least a single strand of DNA. Further, these combinations would be continually irradiated, engulfed with electromagnetic energy, thereby providing these coalescing organic molecules and strands of DNA with additional sources of energy. To speculate: eventually this energized DNA-membranous-protein-plasmoid complex would have begun to function as a proto-organism with all its needs provided by the chemically and electromagnetically enriched surrounding environment. Even if these initial strands of DNA consisted of only four base pairs, these complex plasmodic-cellular structures would have begun evolving. With every replication the plasmoid genome would have expanded and become more variable and more complex.

Some authors have argued that plasmas may represent a form of 'pre-life' or inorganic non-biological life and may have provided an incubating environment for the fashioning of DNA-based life. Plasmas in the thermosphere are also referred to as "dusty plasma" because they have incorporated dust, grains, and the remnants of carbonaceous chondrites that have shattered upon striking the upper atmosphere. The implications are that some upper atmospheric plasmoids likely incorporated the nineteen terrestrial and seventy-three extraterrestrial amino acids that have been detected in carbonaceous chondrites and all five of the nucleobases that make up DNA and RNA. In fact, all the common elements essential to life flow through space and may become incorporated within and beneath the multiple layers of plasma, including carbon, oxygen, phosphorus, sulfur, nitrogen, cyanide, calcium, and hydrogen and which, when irradiated will form organic molecules.

Plasmas, in fact, form spiral helix-like loops when exposed to high intensity currents which act to compress and pinch the layers together. Plasmoids with twisted corkscrew spiraling shapes are commonly observed in the thermosphere and are created in the laboratory when plasmas are exposed to high intensity electric currents. Because dust and fragments from meteorites are likely highly charged, they may trigger the formation of double layers within which this molecular-protein-amino acid complex and the building blocks of nucleotides and other vital prebiotic molecules could have been exogenously incorporated; perhaps forming a nucleus. Moreover, once embedded within the plasma, they may twist, spin, and may form helical structures that can evolve into a double helix similar to the double helix of DNA. Spiral helical internal structures are common among plasmoids in the thermosphere.

We have provided three distinct science-data-based scenarios in which entities of the thermosphere classified as plasmoids/plasmas may have acquired RNA/DNA and a functional genome: (1) via incorporation and assimilation of all the necessary ingredients available in space and the upper atmosphere, (2) via the incorporation and assimilation of the genomes of organisms cast into the upper atmosphere, and 3) via terrestrial genomes which may have been assimilated into the already established genomes of plasmas. If the plasmoids of the thermosphere have acquired RNA, DNA, and a genome, and given their complex behavior, then they should be classified as a "fourth domain of life" and possibly the "first domain" as life may have begun within a plasma.

Plasmas may be pumping electrical energy from one plasma to the next, via twisted filaments and ropes called "Birkeland currents." In consequence; and relying upon the descriptions first employed by Sagan and depending on their size, shape, density, and charge, some plasma in the thermosphere appear to behave as "Hunters" and others as "Grazers" (scavengers) and "Thunderstorm Divers" and those that remain relatively unmoving could be classified as "Floaters."

Solar Plasmodic Electromagnetic Flux "Ropes" and "Cables"

Analysis of NASA space shuttle footage revealed networks of "ropes" "cables" and "chains" completely surrounding an electrified tether. These are likely twisted strands of plasma with a tubular geometry as commonly detected in the magnetosphere; and which transmit electric currents from one region of the magnetosphere to another. The sun emits an electrically conducting plasma which streams radially outward at supersonic speeds of about 500 km/s into interplanetary space thereby producing a solar wind which consists of electrons, protons, and ions, including Helium ions, all of which becomes increasingly dilute as the plasma wind expands radially and with increased distance from the sun. This outflow of plasma is the result of the supersonic expansion of the solar corona and consists of high temperature fast particles, electrons and also ions which reach velocities ranging from 300 km/s and 1500 km/s thereby enabling this ejecta to overcome solar gravitation and escape into interplanetary space. However, even as it flows through space, this plasmodic wind remains electrically highly conducting, acting as a superconductor and forms an interplanetary magnetic field. Once the solar wind traveling at supersonic speed hits Earth's dipole magnetic field, it is deflected and slowed down, and a bow shock wave is generated in front of Earth. In consequence, a substantial fraction of the solar wind particles' kinetic energy is converted into thermal energy whereas the remainder continues to serve as a highly conducting plasma consisting mainly of electrons, protons, ions, and highly charged particles and dust and debris. These highly charged plasmas are distributed in a variety of directions and can form interlinked networks of plasmodic rope-chain-cables that generate electric currents and electromagnetic fields from which globular, cellular plasmas may form and bulge outward. The topography of the solar-terrestrial-atmospheric environment can be discerned, with the sun located to the far lower right of the tether. The solar wind is therefore blowing against the terrestrial-upper atmospheric (geomagnetic) field. Although fragmented cables are apparent in almost all quadrants of this film footage as depicted in published freeze frames, those closest to the tether, are coiled together forming interconnected networks of looping cables. This suggests that the tether is amplifying the charge of the plasma flux ropes.

It is believed that plasmas in space and the thermosphere react to and may form cellular entities in response to a number of interacting forces, including intermittent turbulence, geomagnetic storms, coronal mass ejections, solar flares, eclipses, the waxing and waning of sunlight, and atmospheric waves coupled with ionization. When ionized, the result is a plasma. As predicted by Maxwell's equations and as has been since confirmed, electric fields and powerful stellar magnetic fields are abundant in space. Likewise, cosmic rays are electrically charged. Planets, including Earth also have global electric fields which affect the upper atmosphere as determined by NASA's Endurance mission. Thus, the upper atmosphere and the space around Earth is filled with electrons, protons, and energy created by interactions between the Earth's magnetic field, solar wind, and cosmic rays—as well as dust and debris from shattered meteorites—all of which contribute to the creation and behavior of double layered plasmoids in the thermosphere. Specifically, as charged particles enter Earth's upper atmosphere and strip away electrons from atoms, the remaining protons, neutrons and particles become ionized with a net positive charge which repel, whereas those electrons that have been stripped away have a negative charge and form negative ions; all of which respond to electrical and magnetic fields thereby forming negative and positive double layers that simultaneously repel and attract yet preventing the layers from merging. Thus, plasma in the thermosphere becomes encapsulated with cellular layers. When dust, cosmic debris, and shattered remnants of meteorites are incorporated within a plasma, and if charged, they may trigger the creation of encapsulating double layers and the composite is referred to as a "dusty plasma." In addition, positive ions, which are the major constituent of the solar wind, accelerate outward from the sun and form plasma filaments, ropes, cables and tubes.

Electric currents commonly form twisting plasma filaments; e.g. in lightning, the penumbrae of sunspots, and the sun's photosphere and looping prominences. Galactic and solar electromagnetic impulses coupled with Earth's magnetic field and plasmasphere also interact and produce auroras and donut shaped magnetic filaments that twist, curve, loop and spiral together forming double layered ropes and cables of plasma tubes. These plasmodic circuits of magnetic flux rope-chains are prevalent throughout the upper atmosphere but usually require specialized optical equipment to detect them. This is because these plasma-rope-chain-cables only become "visible" in highly energized environments, such as above the sun and in the upper atmosphere where they appear as multi-colored sheets, filaments, and clouds of plasma, i.e. auroras. Long filaments, cables and ropes of magnetized plasma are also produced by electric current which tend to follow magnetic fields. An electric current is caused by high voltage which creates an electric field. The stronger the current, the stronger the electric field. A strong looping current will produce donut shaped magnetic fields and filaments. These filaments have been characterized as magnetic flux tubes and cables which form tangled loops. The formation of filaments, strings, cables, ropes are common throughout the cosmos and form endless circles and spirals: an infinite helix—all referred to as "field aligned currents." The size of the spiral helix is determined by the strength of the magnetic field—as strength increases and the spiral will tighten forming chains. Portions of the chains may begin to assume cellular forms as they bulge out and "break off." Once a double layered "dusty" plasma is formed, they and the dust and debris internalized, may continue to spiral, forming an internal helix. This may also cause positively and negatively charged particles and debris to diffuse together, creating bright illuminated arc discharges; also known as ambipolar diffusion. When these particles interact—including and especially those within the plasma layers, they generate electric fields which affect their motion. These electromagnetic rope-chains typically have a tubular geometry and have been produced in the laboratory and observed in a variety of astrophysical settings. The most active areas of the sun, for example, produce flux ropes that form chains and loops. These magnetic twisting ropes flow with the solar wind—a function of interactions with the magnetic fields and magnetic clouds produced by coronal mass ejections and are affected by the south-then-north tilt of the magnetic field. In the upper atmosphere of Earth, these rope-chains also interact with charged particles, leading to phenomena like filmy-cloudy auroras, which are in fact twisted strands of plasmas that resemble sheets and filaments of ghostly wavy light. These rope-cable-chains, therefore, are electromagnetic plasmas that produce additional plasmas that can take a variety of shapes and forms, e.g. auroras.

If the plasma ropes in the thermosphere also function like neural networks, and if the same patterns repeat within the interior of a plasmoid, then, those networks might provide the foundation for intelligent behavior. Based on this evidence and eyewitness accounts these complex, communicative and interactive behaviors appear directed and purposeful and under "intelligent control." It is therefore reasonable to hypothesize that some plasmas/plasmoids—especially Hunters and those that descend into the lower atmosphere—are "sentient" "aware" display "intelligence" and constitute a fourth domain of life that dwells primarily in the upper atmospheres of Earth. It has also been documented that plasmoids in the thermosphere (and those generated experimentally) contain internal glowing spheres that are not uncommonly associated with radiating arcs that connect these spheres to the outer membrane layers or to other adjacent internal spheres. It has been speculated that these could be construed as plasmoid genomes or neural ganglia connected to nerve nets, i.e. neural networks, similar to the nervous system of planarians, the common earthworm, and insects.

These electromagnetic-rope-cable networks discovered in the thermosphere, in some respects, resemble and may also function in a manner somewhat similar to the "neural circuity" of the brain which also generates electromagnetic fields. The human nervous system consists of around 80 billion neurons which, when excited, transmit electrified "action potentials" down the length of their axon to a synaptic gap junction. Electricity and/or clouds of chemical neurotransmitters are then released into the gap junction, exciting dendritic receptors that lead to yet another neuron which may transmit the information received to yet another neuron or a host of neurons, via actions potentials propagated along the length of its axons. These neural circuits—vast bundles of insulated cables—enable different brain areas to communicate and share information and contribute to the experience of conscious awareness. Consider also that dusty upper atmospheric plasmoids have likely incorporated not only all the ingredients for fashioning RNA and DNA, but innumerable bacteria, algae, fungi, lichens, plants, insects, fish-eggs—and other organisms and their DNA—lofted into the thermosphere.

Therefore, to speculate: these networks of plasmoid-rope-chains, and the internal ganglia-spheres identified within plasmas, may also function like neural networks and generate and provide an electromagnetic substrate for plasmodic awareness. Many scientists agree: plasma is a fourth state of matter, the highest energy state of matter, and much of the universe, perhaps up to 99.8% consists of plasma. There is also a direct association between the functioning of the brain and plasma physics and the production of electromagnetic waves with estimated frequencies of 0.5 to 100 Hz and with an amplitude of 110 to 4μV. The neurons of the brain interact via electrical impulses—the action potential—which often involve the release of various neurotransmitters at synaptic junctions; i.e. from an axon to dendrite. However, whereas plasma is a form of matter, electricity is not considered to be a "state of matter" but pure energy released by the flow of electrons: an electromagnetic property shared with neurons and lightning which is a plasma that generates transient plasmas, and which attracts plasmoids in the thermosphere that gather above and descend into thunderstorms; whereas the brain radiates electromagnetic waves and shares functional characteristics with plasma physics, and plasma-lightning share basic functional principles with nerve cell activity.

Electricity and lightning in fact shares analogous features with the action potential that travels from the neuron down the length of the axon. In brief: different areas of the brain and linked together and communicate via neurons which send an electrical signal—an action potential—from the neuron's cell body down the length of an appendage, an axon. These positive electrical charges could be likened to a lightning bolt and thus, plasma.

The electromagnetic plasmoid chain-rope-cables in the thermosphere are also punctuated with bulbous spines that may grow and produce, or conversely, attract "Floaters." Lightning and electricity have also been considered essential for the creation of all the necessary elements and nucleotides to create a living organism. Lightning, electromagnetic activity, are linked to the origins of life and nerve cell conduction. This leads us back to self-illuminating plasmas which are electromagnetic entities that are associated with plasmodic rope networks that are charged, and which interact via the release of electrically charged clouds of plasma and dust particles. Plasmas give off light, and it is well established that neural activity can be altered or suppressed by light introduced via optic micro-fibers. Light is not a plasma, but a form of energy; i.e. electromagnetic radiation that travels as waves and is made up of particles called photons. However, all light is produced by plasma, the sun being a primary source of light in this solar system.

The brain is a plasmodic electromagnetic organ which is affected by light and whose action potentials from axons to dendrites share features common to plasma-lightning. It is also these neurons, axons, and dendrites and their associated electromagnetic activity which link and "entangle" the brain to the electromagnetic quantum continuum which, some believe, is consciousness in its totality. Plasmas are electromagnetic entities and may possess a virtual plasma-based nervous system that is not "physical" or "hard wired" but electromagnetic and whose nerve network and circuitry is electric but may have incorporated charged particles, dust, and debris, thereby creating stable interconnected electric circuits and ganglia that function in a manner similar to the eukaryotic nervous system; and which provides the foundation for intelligence and purposeful behavior. However, a nervous system is not a perquisite for life; and even plants—which are devoid of axons, dendrites and neurons—display behavior and an awareness of their environment.

Plasmas in the thermosphere also appear to display patterns common on a solar and galactic level; e.g. comets, colliding galaxies, and perhaps pulsars in respect to plasmodic oscillations in dim vs bright light: except that pulsars spin whereas these oscillating plasmas shrink and expand. There is now abundant observational evidence that magnetic flux filamentation occurs and is common at all astrophysical levels and that generates structures on smaller and smaller scales from galaxies to stars to planets, and perhaps to the scale of the nervous system. Magnetized plasma ropes, cables, and filaments have been observed at the scale of superclusters interconnecting multiple galaxies, within clusters, along galactic spiral arms, within star forming molecular clouds, within solar system, above the sun, in the magnetosphere/plasmasphere and in the thermosphere as documented in this report. Plasmas in the thermosphere pulsate with light. Red giants also pulsate (e.g. Betelgeuse, Mira A), as do binary stars and "variable stars" which may have regular or irregular pulsating patterns—possibly because there are electrical discharges between nearby stars. Plasmas (Floaters) form semi-conglomerates of semi-stationary plasmoids grouped together, or in which one within a grouping will orbit around and make contact with the others; and/or which are interconnected with luminous clouds of plasma. Such grouping and behavior are also common at astronomical levels. And then there are the thousands of galaxies that appear to be colliding. Plasmas also collide. Arp has documented and photographed numerous physical connections between galaxies, including those with high vs low redshift values—connected by a luminous bridge or tail—as well as hundreds of galaxies that appear to be interacting. Plasmas streak at varying velocities across the thermosphere, targeting and colliding with other plasmas and often leaving a plasma-comet-like tail in their wake; whereas comet luminosity has been attributed to electrical excitement. Be they large or small, comets frequently display "non-gravitational" erratic motions, as do plasma Hunters in the thermosphere. And comets produce filamented plasma tails that may stretch for tens of millions of kilometers across the solar system whereas plasmas in the thermosphere produce filamented plasma tails that may stretch tens of kilometers

This report provides examples of plasma mitosis, the ejection of clouds of plasma prior to and after collisions between plasmoids and the ejection of mini-plasmas; processes that appear to represent a form of reproduction and subsequent growth. Hence, although it is not possible to demonstrate that all seven characteristics of life are present, this report provides sufficient information to conclude that plasmoids likely constitute a 'form of life.'

It is widely accepted that plasma constitute a 'fourth state of matter' whereas Alfven considered plasmas to be the "first state of matter." As detailed in this article, plasmas in the thermosphere engage in complex behavior and interactions that appear purposeful and under intelligent control.

Therefore, plasmas may be alive and may be sentient and may represent a fourth domain, or even a first domain of life. In this solar system, in the atmosphere above Earth, it can be assumed that the plasmasphere (magnetosphere) has been in existence for at least 3.5 billion years following the great bombardment of meteors, asteroids, comets, and oceans of ice. We have hypothesized that over the ensuing billions of years, that dusty plasmas incorporated the shattered remnants of meteors and all the necessary ingredients for the fashioning of RNA and DNA; and later incorporated the genomes of living creatures that were cast into the upper atmosphere by powerful winds, hurricanes, tornadoes and bolide impacts. These theoretical scenarios give rise to at least four hypotheses: (1) Dusty plasmas are the first domain of life and incubated and synthesized RNA then DNA thus leading to the origins of life. (2) Dusty Plasma acquired RNA/DNA after incorporating and incapsulating innumerable living organisms that were cast into the upper atmosphere. (3) Dusty plasmas were the first to synthesis RNA and DNA, and upon acquiring and encapsulating organisms propelled into the upper atmosphere, incorporated their DNA into the plasma DNA via horizontal gene transfer. (4) Plasmas (plasmoids) are a form of 'pre-life' or an inorganic non-biological form of 'life" which function according to the same principles of electromagnetism which govern the functioning of living organisms be they plants or organisms whose behavior is controlled by a brain. We have also hypothesized that plasmoids—hunters in particular—may have developed a virtual plasma-based nervous system. The brain of eukaryotes is an electromagnetic organ and the nervous system functions according to and is governed by electromagnetic-plasma principles. However, we are not proposing that plasmoid have a "physical" "hard wired" nervous system, but an electromagnetic nervous system whose nerve network and circuitry is electric but which may have incorporated charged particles, dust, and debris, thereby creating stable interconnected electric circuits and ganglia that function in a manner similar to the eukaryotic nervous system; and which provides the foundation for sentience, intelligence, and consciousness. The fact is, as documented in this and earlier reports, plasmas in the thermosphere engage in complex behaviors and interactions. These include what may be communication via a language of light with some plasmoids rapidly oscillating in brightness and size as they approach vs pass by other plasmas. Then there are the ejections of clouds of plasma in their wake, the formation of plasma bridges between one or more plasma, plasma mitosis, and the secretion of "emissary" or "messenger" plasmas that detach (repulse) and may then be drawn toward the plasma that triggered the fissioning. This same principle may apply when two plasmas target and crash into one another, or when plasmas merge and separate. However, each of these plasmas may have different voltages, density, temperatures and chemistry which may rapidly change, and all of which generates electric currents and charge separation especially when they crash through and pierce each other. Therefore, all these interactions and complex behaviors may be entirely due to differential electromagnetic activity and negative vs positive electrical charges and push-pull magnetic repulsion and attraction as similar interactions have been observed between galaxies, pulsars, binary stars and red giants. Hence, if the plasmas in the thermosphere are alive, then, could it be said that the plasma universe, the electromagnetic quantum continuum, is also alive? We cannot provide definitive conclusive answers to any of these questions other than to conclude that the data presented challenge all conceptions of what constitutes "life" and how life began.

Is God Plasma?

by Roberta Ruth Hill & Brooke West

"A person hears only what he can understand."
Goethe

Plasma is the fourth state of matter and makes up about 99% of the universe. We are surrounded by plasma and yet it is mostly invisible until it interacts with the electromagnetic spectrum. Scientific experiments have shown that plasma holds the blueprint for microscopic cells and the spiraling DNA. It has been shown in recent experiments that dust in plasma starts spiraling like DNA and the galaxies in our universe. Plasma seems to be the precursor for all life as we know it, and it seems to have all of the attributes that are given to a creator God and the ancient gods. Could the spirit of God be plasma? We attempt to answer that question in this book.

> "And the Lord God formed man of the dust of the ground and breathed into his nostrils the breath of life; and man became a living soul." *Genesis* 2:7

It is apparent that plasma is capable of building a cell-like structure, which consists of a double layer membrane in which it transports negative and positive ions to create energy in the form of electricity. It is even more surprising to realize that our very own Earth has a plasma double layer in the ionosphere, and that it creates the Birkeland currents which feed the electromagnetic displays of the colorful aurora at the Earth's poles. We can furthermore make a connection that the two double layers are actually the same thing, and that we ourselves are also made up of thousands of double layer cells within our own bodies. The plasma spheres that were created in the laboratory show us that this blueprint for all life was started by this very ancient substance that was here in the beginning of the universe and at the beginning of the Earth.

Plasma not only seems to be the fourth state of matter, but it seems to be the blueprint for all matter, as well. We can actually see that plasma double layers are seen throughout the universe from the microscopic to the macroscopic cosmos and that the hermetic principle of "as above, so below" applies to plasma and the entire universe. Plasma seems to be a construct from the spiritual universe to apply a foundation or schematics for building the physical universe, therefore plasma is a prerequisite for all life and the entire structure of our universe. It is the underlying fabric of space time, and from this fabric, all of the patterns of the universe are cut and sewn together. In my chapter on *The Elect*, I put forth the theory that God, the gods and Angels could be some kind of plasma life forms.

I showed scientific proof in my *Biology and Plasma* chapter that plasma seems to mimic many of the characteristics for life. The plasma spheres created in the laboratory showed that plasma is capable of reproducing itself, which makes it able to grow its population. Furthermore, plasma spheres are able to grow in their individual size and are also able to communicate with one another through sound frequencies. Any form of communication denotes intelligence on some level and is evidence that plasma has a sentient quality. The experiment went on to show that plasma was able to form cell-like structures that were eerily similar to our own biological cells. The scientists felt that these plasma spheres or cells were the precursor for organic biological cells, which suggested that they were the original blueprint of all cells and life as we know it. In my *Thunder and Lightning* chapter I proved through specific verses in the *Bible* that lightning was associated with God, the Angels, Jesus, Satan and the beings that Ezekiel observed in the wheels within wheels vehicle. Since lightning is a form of plasma, those particular verses I mentioned in that chapter prove a plasma connection to God. Lightning was also a major component involved with the throne of God that is mentioned in *Revelation* 4, and it appeared to be involved in the fiery serpent pole of *Numbers* 21, as well as the arc of electricity that was made by the Ark of the Covenant. Furthermore, Hutchinson's experiment to recreate the electrical arc in between the Cherubim or electrodes shows definite proof that it is possible that the Ark of the Covenant was an electrical device.

In regards to the gods that are mentioned all over the world, there are specific gods that were said to display lightning or to use lightning as a weapon of some sort. One of the most popular gods that is known worldwide in our modern-day culture is the god of Zeus. His identity and symbol is most often reflected as being a thunderbolt. This shows a connection between the god Zeus and plasma, as well as it shows that he had the ability to use and create lightning for his own purpose. This ability denotes that Zeus could have been a plasma being that interacted with electromagnetic fields to create his fiery lightning bolts. There are other gods that are mentioned from around the world that also had the ability to create lightning.

In the chapter on the *Ark of the Covenant*, I showed more proof to connect the Ark of the Covenant to plasma and showed that it is scientifically possible that the ark could have been used as a communication device. Plasma speakers are able to project sound and music just as well as electromagnetic speakers, and it makes a shape that could very well be described as a whirlwind shape. The videos that show plasma speakers are evidence that God could talk to man through a whirlwind as some verses do suggest. In the biology chapter, I was able to expand on the blueprint of the Ark of the Covenant to show that this pattern also very much duplicates the double layer cell membrane of a plasma sphere, the ionosphere and the biological cell.

In the *Electricity* chapter, I went into detail about how electricity seems to be a major component by which the Israelites were interacting with God. Since plasma interacts with the electromagnetic field, electricity would be a byproduct of that interaction; therefore, electricity proves that plasma is involved with the communication to God. I also pointed out that Ed Leedskalnin, who is the builder of the mysterious Coral Castle had possibly decoded the information of *Revelation* 4 and used it as his blueprint for making his magnetic flywheel to generate free electricity. It also seems as if Tesla was aware of these blueprints, as well as the electromagnetic frequencies that were encoded into the *Bible*.

In the biology chapter, I also talked about the experiment with dust and plasma and how this experiment showed that dust will start forming crystals and spiral shapes. This experiment confirms a statement made by God in the *Bible*, who proclaimed that he made man from the dust of the ground. The fact that dust inside of plasma makes the same shape as our DNA, which are the building blocks for the human body and contains all the blueprints and functions that are needed to make a human being seems to confirm that plasma made human beings. The experiment seems to scientifically explain the creation of human beings as being related to the dust being enveloped and shaped by plasma into the genetics that became the basis for the genesis of mankind. We can clearly see a connection between plasma and God once again.

I believe that I have proven that there is much scientific and mythical evidence that God and the gods are made of plasma and that this best explains the non-physical description of God being a spirit rather than a physical being. God is said to be everywhere and to be omnipotent, and plasma is known to be everywhere in the universe, as well as being a powerful substance with the ability to transform dust to its will and to even control the electrons of atoms. Plasma best explains God's spiritual nature and it best explains our electromagnetic body, which we call the soul.

As *1 Corinthians* 15:44 says, "It is sown a natural body; it is raised a spiritual body. There is a natural body, and there is a spiritual body." Every life form, including the living bio-organisms of the planets with life, has both a physical carbon-based body and a spiritual plasma based body. This knowledge that has been contained within ancient religious books and beliefs from around the world and has been proven in our modern times through scientific studies, research, and experiments.

After all of my research on plasma and ions, I would have to say that I am convinced that all of the evidence certainly proves that God is plasma without a doubt. I would encourage the reader to research this topic thoroughly for themselves and to read more about these experiments that are mentioned in this book. We seemed to have reached the golden age of science, since scientific experiments are now explaining the ancient myths, religious books, and beliefs which are indeed confirming that this ancient knowledge was accurately explaining the very nature of plasma. Everything that has been attributed to God or the gods can be explained thoroughly through the science of plasma. Studying plasma is the key to better understanding the Creator and to better understanding ourselves.

Is God plasma? The evidence supports the hypothesis that God is indeed plasma. There is enough scientific proof that supports this as a legitimate theory that needs to be taken seriously by both scientists and religious scholars alike. Now is the time that science and religion can actually be combined to explain the creation of the universe and it is perhaps surprising that both are actually starting to say the same kind of things now. This is an astonishing truth that needs to be fully recognized by both communities. There no longer needs to be competition by these two organized doctrines, as both science and religion are now becoming the same thing and should be working together in a cooperative effort. Science and religion have both sought to explain our creation for centuries and they no longer must be at odds with one another. Plasma fully explains the manipulation of dust to make physical objects such as biological cells, DNA, stars, and planets. It has been proven by scientists that plasma is a precursor for all life as we know it and that it possibly carries the blueprints for creation. Plasma is more than just the fourth state of matter; it is the substance from which everything is made. Plasma is the underlying fabric of the universe and from that fabric we were all cut and made, like a tailor cuts the pattern of cloth to make a suit. Plasma is a creative force which is the maker of everything in our universe, so in that regard....... God is plasma.

Chariots of Fire—Encounters with Plasma

Ezekiel 1

1 Now it came to pass in the thirtieth year, in the fourth month, in the fifth day of the month, as I was among the captives by the river of Chebar, that the heavens were opened, and I saw visions of God.

4 And I looked, and, behold, *a whirlwind came out of the north, a great cloud, and a fire enfolding itself, and a brightness was about it*, and out of the midst thereof as the color of amber, out of the midst of the fire.

5 Also out of the midst thereof came the likeness of four living creatures. And this was their appearance; they had the likeness of a man.

6 And every one had four faces, and every one had four wings.

7 And their feet were straight feet; and the sole of their feet was like the sole of a calf's foot: and they sparkled like the color of burnished brass.

8 And they had the hands of a man under their wings on their four sides; and they four had their faces and their wings.

9 Their wings were joined one to another; they turned not when they went; they went every one straight forward.

10 As for the likeness of their faces, they four had the face of a man, and the face of a lion, on the right side: and they four had the face of an ox on the left side; they four also had the face of an eagle.

11 Thus were their faces: and their wings were stretched upward; two wings of every one were joined one to another, and two covered their bodies.

12 And they went every one straight forward: whither the spirit was to go, they went; and they turned not when they went.

13 As for the likeness of the living creatures, *their appearance was like burning coals of fire, and like the appearance of lamps: it went up and down among the living creatures; and the fire was bright, and out of the fire went forth lightning.*

14 *And the living creatures ran and returned as the appearance of a flash of lightning.*

15 Now as I beheld the living creatures, behold one wheel upon the earth by the living creatures, with his four faces.

16 The appearance of the wheels and their work was like unto the color of a beryl: and they four had one likeness: and their appearance and their work was as it were *a wheel in the middle of a wheel.*

17 When they went, they went upon their four sides: and they turned not when they went.

18 *As for their rings, they were so high that they were dreadful; and their rings were full of eyes round about them four.*

19 And when the living creatures went, the wheels went by them: *and when the living creatures were lifted up from the earth, the wheels were lifted up.*

20 *Whithersoever the spirit was to go, they went, thither was their spirit to go; and the wheels were lifted up over against them: for the spirit of the living creature was in the wheels.*

21 When those went, these went; and when those stood, these stood; and when those were lifted up from the earth, the

wheels were lifted up over against them: for the spirit of the living creature was in the wheels.

22 And the likeness of the firmament upon the heads of the living creature was as the color of the terrible crystal, stretched forth over their heads above.

23 And under the firmament were their wings straight, the one toward the other: every one had two, which covered on this side, and every one had two, which covered on that side, their bodies.

24 And when they went, I heard the noise of their wings, like the noise of great waters, as the voice of the Almighty, the voice of speech, as the noise of an host: when they stood, they let down their wings.

25 And there was a voice from the firmament that was over their heads, when they stood, and had let down their wings.

26 And above the firmament that was over their heads was the likeness of a throne, as the appearance of a sapphire stone: and upon the likeness of the throne was the likeness as the appearance of a man above upon it.

27 And I saw as the color of amber, as the appearance of fire round about within it, from the appearance of his loins even upward, and from the appearance of his loins even downward, I saw as it were the appearance of fire, and it had brightness round about.

28 As the appearance of the bow that is in the cloud in the day of rain, so was the appearance of the brightness round about. This was the appearance of the likeness of the glory of the Lord. And when I saw it, I fell upon my face, and I heard a voice of one that spake.

The Ophanim

"As I looked, Thrones were set in place, and the Ancient of Days took his seat. His clothing was as white as snow; the hair of his head was white like wool. His throne was flaming with fire, and its wheels were all ablaze."

Daniel 7:9

The ophanim (also called spheres, wheels, whirlwinds) refer to the wheels seen in Ezekiel's vision of the chariot (Hebrew merkabah) in *Ezekiel* 1:15–21. One of the *Dead Sea Scrolls* construes them as angels; late sections of the *Book of Enoch* (61:10, 71:7) portray them as a class of celestial beings who (along with the Cherubim and Seraphim) never sleep but guard the throne of God. In some systems of Christian angelology, they are one of the choirs of angels and are also identified as Thrones.

These "wheels" have been associated with *Daniel* 7:9, the four, eye-covered wheels (each composed of two nested wheels), that move next to the winged Cherubim, beneath the Throne of God. The four wheels move with the Cherubim because the spirit of the Cherubim is in them. The late *Second Book of Enoch* (20:1, 21:1) also referred to them as the "many-eyed ones."

The *First Book of Enoch* (71.7) seems to imply that the Ophanim are equated to the "Thrones" in Christianity when it lists them all together, in order: "...round about were Seraphim, Cherubim, and Ophanim."

It is said that they were the actual wheels of the Lord's Heavenly Chariot (Merkabah). "The four wheels had rims and they had spokes, and their rims were full of eyes round about." They are also frequently referred to as "many-eyed ones."

De Coelesti Hierarchia refers to the Thrones from the *Old Testament* description as the third Order of the first sphere, the other two superior orders being the Cherubim and Seraphim.

The name of the most glorious and exalted Thrones denotes that which is exempt from and untainted by any base and earthly thing, and the super mundane ascent up the steep. For these have no part in that which is lowest, but dwell in fullest power, immovably and perfectly established in the Most High, and receive the Divine Immanence above all passion and matter, and manifest God, being attentively open to divine participations.

Ophanim or Thrones

The Book of Enoch, From: *The Apocrypha and Pseudepigrapha of the Old Testament*, by R.H. Charles, Section I, Chapters I-XXXVI

And the vision was shown to me thus: Behold, in the vision, clouds invited me and a mist summoned me, and the course of the stars and the lightnings sped and hastened me, and the winds in the vision caused me to fly and lifted me upward, and bore me into heaven. And I went in till I drew nigh to a wall which is built of crystals and surrounded by tongues of fire: and it began to affright me. And I went into the tongues of fire and drew nigh to a large house which was built of crystals: and the walls of the house were like a tessellated floor made of crystals, and its groundwork was of crystal. Its ceiling was like the path of the stars and the lightnings, and between them were fiery cherubim, and their heaven was clear as water. A flaming fire surrounded the walls, and its portals blazed with fire. And I entered into that house, and it was hot as fire and cold as ice: there were no delights of life therein: fear covered me, and trembling got hold upon me. And as I quaked and trembled, I fell upon my face. And I beheld a vision, And lo! there was a second house, greater than the former, and the entire portal stood open before me, and it was built of flames of fire. And in every respect it so excelled in splendor and magnificence and extent that I cannot describe to you its splendor and its extent. And its floor was of fire, and above it were lightnings and the path of the stars, and its ceiling also was flaming fire. And I looked and saw therein a lofty throne: its appearance was as crystal, and the wheels thereof as the shining sun, and there was the vision of cherubim. And from underneath the throne came streams of flaming fire so that I could not look thereon. And the Great Glory sat thereon, and His raiment shone more brightly than the sun and was whiter than any snow. None of the angels could enter and could behold His face by reason of the magnificence and glory and no flesh could behold Him. The flaming fire was round about Him, and a great fire stood before Him, and none around could draw nigh Him: ten thousand times ten thousand stood before Him, yet He needed no counselor. And the most holy ones who were nigh to Him did not leave by night nor depart from Him. And until then I had been prostrate on my face, trembling: and the Lord called me with His own mouth, and said to me: "Come hither, Enoch, and hear my word." And one of the holy ones came to me and waked me, and He made me rise up and approach the door: and I bowed my face downwards.

2 Enoch 20: 1

In the first month, on the assigned day of the first month, I was in my house alone. And I lay on my bad sleeping. And, while I slept, a great distress entered my heart, and I was weeping with my eyes in a dream. And I could not figure out what this distress might be, |nor| what might be happening to me. Then two huge men appeared to me, the like of which I had never seen on earth.

Their faces were like the shining sun;
their eyes were like burning lamps;
from their mouths fire was coming forth;
their clothing was various singing;
their wings were more glistening than gold;
their hands were whiter than snow.

2 Enoch 21: 1

About how here the angels left Enoch at the edge of the 7th heaven, and departed from him invisibly. And they do not leave by night, nor depart by day, standing in front of the face of the LORD, and carrying out his will—cherubim and seraphim standing all around his throne, six-winged and many-eyed; |and| they cover his entire throne, singing with gentle voice, in front of the face of the LORD. Holy, Holy, Holy, Lord Sabaoth, Heaven and earth are full of his glory.

The Book of Enoch, Chapter 71

1. And it came to pass after this that my spirit was translated And it ascended into the heavens: And I saw the holy sons of God. They were stepping on flames of fire: Their garments were white [and their raiment], And their faces shone like snow.

2. And I saw two streams of fire, And the light of that fire shone like hyacinth, And I fell on my face before the Lord of Spirits.

3. And the angel Michael [one of the archangels] seized me by my right hand, And lifted me up and led me forth into all the secrets, And he showed me all the secrets of righteousness.

4. And he showed me all the secrets of the ends of the heaven, And all the chambers of all the stars, and all the luminaries, Whence they proceed before the face of the holy ones.

5. And he translated my spirit into the heaven of heavens, And I saw there as it were a structure built of crystals, And between those crystals tongues of living fire.

6. And my spirit saw the girdle which girt that house of fire, And on its four sides were streams full of living fire, And they girt that house.

7. And round about were Seraphin, Cherubic, and Ophannin: And these are they who sleep not And guard the throne of His glory.

8. And I saw angels who could not be counted, A thousand thousands, and ten thousand times ten thousand, Encircling that house. And Michael, and Raphael, and Gabriel, and Phanuel, And the holy angels who are above the heavens, Go in and out of that house.

Psalm 18: 6-16

In my distress I called upon the Lord, and cried unto my God: he heard my voice out of his temple, and my cry came before him, even into his ears.

Then the earth shook and trembled; the foundations also of the hills moved and were shaken, because he was wroth.

There went up a smoke out of his nostrils, and fire out of his mouth devoured: coals were kindled by it.

He bowed the heavens also and came down: and darkness was under his feet.

And he rode upon a cherub, and did fly: yea, he did fly upon the wings of the wind.

He made darkness his secret place; his pavilion round about him were dark waters and thick clouds of the skies.

At the brightness that was before him his thick clouds passed, hail stones and coals of fire.

The Lord also thundered in the heavens, and the Highest gave his voice; hail stones and coals of fire.

Yea, he sent out his arrows, and scattered them; and he shot out lightning, and discomfited them.

Then the channels of waters were seen, and the foundations of the world were discovered at thy rebuke, O Lord, at the blast of the breath of thy nostrils.

He sent from above, he took me, he drew me out of many waters.

Merkabah, the Chariot of God, the Chariot of Creation

Merkabah mysticism or Chariot mysticism, is a school of early Jewish mysticism, c. 100 BCE—1000 CE, centered on visions such as those found in *Ezekiel 1* or in the *hekhalot* literature ("palaces" literature), concerning stories of ascents to the heavenly palaces and the Throne of God, later described by Saint John in the book of *Revelation*.

The Hebrew term Merkavah relates to the throne-chariot of God in prophetic visions. It is most closely associated with Ezekiel's vision of the four-wheeled vehicle driven by four *hayyot* "living creatures", each of which has four wings and the four faces of a man, lion, ox, and eagle.

According to the verses in Ezekiel and its attendant commentaries, his vision consists of a chariot made of many heavenly beings driven by the "Likeness of a Man." The base structure of the chariot is composed of four beings. These beings are called the "living creatures." The bodies of the creatures are "like that of a human being", but each of them has four faces, corresponding to the four directions the chariot can go. The faces are that of a man, a lion, an ox (later changed to a Cherub in *Ezekiel* 10:14) and an eagle.

Each of the *hayyot* angels also has four wings. Two of these wings spread across the length of the chariot and connect with the wings of the angel on the other side. This creates a sort of 'box' of wings that forms the perimeter of the chariot. With the remaining two wings, each angel covers its own body. Below, but not attached to, the feet of the *hayyot* angels are other angels that are shaped like wheels. These wheel angels, which are described as "a wheel inside of a wheel", are called Ophanim. These wheels are not directly under the chariot but are nearby and along its perimeter. The angel with the face of the man is always on the east side and looks up at the "Likeness of a Man" that drives the chariot. The "Likeness of a Man" sits on a Throne made of sapphire.

The *Bible* later makes mention of a third type of angel found in the Merkabah called "Seraphim" (lit. burning) angels. These angels appear like flashes of fire continuously ascending and descending. These Seraphim angels power the movement of the chariot. The chariot is in a constant state of motion, and the energy behind this movement runs according to these beings. The movement of the Ophanim is controlled by the "Living creatures", or *Hayyot*, while the movement of the *Hayyot* is controlled by the Seraphim. The movement of all the angels of the chariot is controlled by the "Likeness of a Man" on the Throne.

Vimana, Chariots and Flying Palaces

Vimana are mythological flying palaces or chariots described in Hindu texts and Sanskrit epics. In the Ramayana, the famous Pushpaka vimana is depicted three times, twice flying in the sky and once landed on the ground and is described as follows:

> "The Pushpaka Vimana that resembles the Sun and belongs to my brother was brought by the powerful Ravana; that aerial and excellent Vimana going everywhere at will ... that chariot resembling a bright cloud in the sky ... and King Rama got in, and the excellent chariot at the command of the Raghira, rose up into the higher atmosphere."

This is the first flying vimana mentioned in existing Hindu texts (as distinct from the gods' flying horse-drawn chariots). Pushpaka was originally made by Vishvakarma for Brahma, the Hindu god of creation.

In the Mahabharata a title of Uparichara was received by a king named Vasu after a Vimana flying chariot was granted to him by Indra, who was pleased with him. This chariot enabled him to wander above all mortals. The Vimana appeared as a crystalline entity in the sky and is described in the *Mahabharata*—Volume 1—*Adi Parva*—064:

> "The celestial chariot, enjoyable to the deities and beautifully crystalline in the sky, that I have graciously bestowed upon you will come to you. You alone, on the splendid chariot, residing above all mortals, will wander above, like the embodiment of the divine."

The Chaldean Oracles

There is above the Celestial Lights an Incorruptible Flame always sparkling; the Spring of Life, the Formation of all Beings, the Original of all things! This Flame produceth all things, and nothing perisheth but what it consumeth. It maketh Itself known by Itself. This Fire cannot be contained in any Place, it is without Body and without Matter. It encompasseth the Heavens. And there goeth out from it little

Sparks, which make all the Fires of the Sun, of the Moon, and of the Stars. Behold! what I know of God! Strive not to know more of Him, for that is beyond thy capacity, how wise soever thou art. As to the rest, know that unjust and wicked Man cannot hide himself from the Presence of God! No subtilty nor excuse can disguise anything front His piercing Eyes. All is full of God, and God is in All.

There is in God an Immense Profundity of Flame! Nevertheless, the Heart should not fear to approach this Adorable Fire, or to be touched by it; it will never be consumed by this sweet Fire, whose mild and Tranquil Heat maketh the Binding, the Harmony, and the Duration of the World. Nothing subsisteth but by this Fire, which is God Himself. No Person begat Him; He is without Mother; He knoweth all things, and can be taught nothing. He is Infallible in His designs, and His name is unspeakable, Behold now, what God is! As for us who are His messengers, We are but a Little Part of God.

On Matter: the World and Nature

The Matrix containing all things.

Wholly divisible, and yet indivisible.

Thence abundantly springeth forth the generations of multifarious Matter.

These frame atoms, sensible forms, corporeal bodies, and things destined to matter.

The Nymphs of the Fountains, and all the Water Spirits, and terrestrial, aërial and astral forms, are the Lunar Riders and Rulers of all Matter, the Celestial, the Starry, and that which lieth in the Abysses.

According to the Oracles, Evil is more feeble than Non-entity.

We learn that Matter pervadeth the whole world, as the Gods also assert.

All Divine Natures are incorporeal, but bodies are bound to them for your sakes. Bodies not being able to contain incorporeals, by reason of the Corporeal Nature, in which ye are concentrated.

For the Paternal Self-begotten Mind, understanding His works sowed in all, the fiery bonds of love, that all things might continue loving for an infinite time. That the connected series of things might intellectually remain in the Light of the Father; that the elements of the World might continue their course in mutual attraction.

The Maker of all things, self-operating, framed the World. And there was a certain Mass of Fire: all these things Self-Operating He produced, that the Body of the Universe might be conformed, that the World might be manifest, and not appear membranous,

For He assimilateth the images to himself, casting them around his own form.

For they are an imitation of his Mind, but that which is fabricated hath something of Body.

There is a Venerable Name, with a sleepless revolution, leaping forth into the worlds, through the rapid tones of the Father.

The Ethers of the Elements therefore are there.

The Oracles assert that the types of Characters, and of other Divine visions appear in the Ether (or Astral Light).

In this the things without figure are figured.

The Ineffable and Effable impressions of the World.

The Light hating World, and the winding currents by which many are drawn down.

He maketh the whole World of Fire, Air, Water, and Earth, and of the all-nourishing Ether.

The Orphic Hymns

In Orphic cosmogony, Phanes is often equated with Eros, Ether, Dionysos, Eubouleus, Antauges, or Mithras and has been depicted as a deity emerging from a cosmic egg entwined with a serpent: the Orphic egg. He had a helmet and had broad, golden wings.

To Protogonos

Upon two-natured Protogonos, great and ether-tossed, I call; born of the egg, delighting in his golden wings, the begetter of blessed gods and mortal men; he bellows like a bull.

Erikepaios, seed unforgettable, attended by many rites, ineffable, hidden, brilliant scion, forever in whirring motion, you scattered the dark mist, the mist that lay before your eyes.

Flapping your wings, you whirled about throughout this world, you brought pure light.

For this I call you Phanes, yes, I call you Lord Priapos, I call you bright-eyed Antauges.

O blessed one of the many counsels and of the many seeds, come joyfully to the celebrants of this holy rite, of this very intricate rite.

To Ether

Yours are Zeus' lofty dwelling, endless power, too; of the stars, of the sun, and of the moon you claim a share.

O tamer of all, O fire-breather, O life's spark for every creature, sublime Ether, best cosmic element, radiant, luminous, starlit offspring,

I call upon you and I beseech you to be temperate and clear.

Plasma and the Seven Ethers

Plasma defies all of the standard laws of physics, including: gravity, the weak and strong nuclear force, electricity, and any other standard scientific approach to plasma as an "atomic substance."

Plasma, in its many forms, cannot be understood by a modern mind steeped in scientific materialism (scientism). Plasma defies every known law that scientism has hypothesized in their ever-changing theories. Plasma defies gravity and scientism hasn't even "discovered" levity yet. Plasma is electro-static, electro-magnetic, electro-gravitational, and is a magnetohydrodynamic generator that confounds scientists in their efforts to create a plasma generator that can hold the fusion power of the Sun in a containment vessel on Earth. Scientism is looking in all the wrong places, but it has utilized the forces, effects, and power of plasma in medicine, weapons, and power generation and yet have spent little time looking for plasma in the human body or its natural conditions. Scientists have looked for plasma everywhere except in their own hearts, which is a plasma magnetohydrodynamic generator that produces a large torus electro-magnetic field allowing the other two generators in the head and in the gut to operate in an invisible electromagnetic field. Plasma is the miracle of creation, the science of the future, and the Tree of Life that is found in the human body, the Earth as a whole, and the plasma-filled cosmos.

We presented this book as a three-phase introduction to the topic of plasma: outer knowledge, inner wisdom, and secret teachings. Just like the ancient Eastern spiritual practices had phases that an aspirant had to progress through, so too the deeper wisdom and secrets of plasma have yet to be disclosed to the uninitiated. The outer knowledge of plasma grows each day with new inventions and uses for plasma, in its many forms. Of course, scientism first made weapons of plasma, even though they did not understand what they were working with, just like electromagnetism and nuclear energy. It is because of this lack of morality that Rudolf Steiner said in his time that humans were not moral enough to hear and understand the details of the "critical phenomena of life", or of the life giving forces and beings who created it. If scientism found that plasma was "alive", they would immediately dissect and kill it by "studying" it, as they always do. Therefore, we have presented in the first third of this book the data and facts we accumulated from scientism and presented them as a foundation upon which to build deeper secrets.

The inner wisdom of plasma has been known since the beginning of human consciousness, which, in ancient times was also accompanied by a simple and natural clairvoyance that "witnessed" plasma in lightning, ball lightning, descending plasmoids, and particularly in the "electric" body of the human being centered in the nerve and blood systems of the body. Ancient clairvoyants knew that the human heart, head, and gut have plasmic energy flowing out of them and into the entire cardio-vascular, metabolic, and nervous systems of the body. These energies are now invisible to modern materialists, even though a new clairvoyance is developing the capacities to see plasma, and her sisters, the seven ethers. There are endless stories of ancient peoples encountering things like: "a burning bush", a "ball of fire", the aurora borealis, angel messengers of gods, or gods like Hermes, Phanes, Mithra, Aditi and Aditya first born, Eros, Zeus and his lightning bolts, Prometheus the fire bringer, Helios on his fiery chariot, etc., etc., etc. There were endless names and descriptions of the forces and beings encountered by the ancients. They also could see the energy, chi, bio-electricity moving through the chakras and the meridians of the human body. Life was filled with the sacred effects of plasma. We can learn much from these accounts, which were often attributed to the gods.

As a born clairvoyant, I witnessed these forces, beings, and effects all around me and in others but never had words to describe them; or when I tried, I was assumed to be crazy or making up nonsense. As a child, I had studies scientism through reading the *World Book Encyclopedia* and found nothing in scientism that explained what I encountered continuously. I spent my youth studying cybernetics, bio-electricity, chi, healing modalities that tried to align with these forces, mitochondrial production of energy in the human cell, atmospheric forces, ancient teachings on Qigong, Taoist methods of manipulating energy in the body, and many other avenues that still did not create a complete picture of what I was perceiving. Then, I came across Rudolf Steiner's etheric formative forces and plasma. Eventually, I understood that these two sets of forces (spheres of activity, waves, frequencies, creative forces, and entities) were polar opposite forces working in realms that interpenetrate the physical, material world. I then understood that matter is not substantial, it is temporary and changing dynamically all the time, evolving and metamorphosing into new forms. I realized that I didn't have thoughts that could encompass what I witnessed, nor words to describe these actions of spirit manifesting in material life.

Eventually, I simply became a healer who worked with these bio-electric forces and beings that were invisible to others but clearly perceived by me. The desire to understand these energies led me outdoors to observe and study thunderstorms and the many tornadoes that occurred where I lived in Missouri. I often observed balls of lightning, both in clouds and descending close to the Earth.

Once, when I was young, I witnessed a fantastic summer thunderstorm that created hundreds of lightning strikes and non-stop thunder. The thunder and lightning was so intense that my family took shelter in the house and hid in terror and fear. I stood outside and witnessed the storm with great pleasure and exuberance. I had always dreamt of being struck by lightning and being very happy about it. I believed that those types of "translation into heaven" were experienced by Ezekiel, Enoch, Melchizedek, and Elijah, among others. I often climbed on top of roofs to get as close as I could to the clouds where ball lightning was roiling around like a fiery ball of energy.

After watching the storm advance straight at me for more than an hour, all lightning and thunder stopped abruptly. I could see a large ball of lightning rolling around the front of the thunderhead which had finally reached my position. Then, the ball lightning slowly dropped from the cloud towards the ground about a quarter of a mile from where I was standing. It began to move quicker, skimming over the outline of the houses it was flying over. It was only a couple of feet from the roofs of the houses but did not touch them, just outlined their silhouette. The fiery ball roiled over one house after the next and continued on a straight course directly at me. I realized that the fire-ball would soon be on my own street flying over the house next to mine. I felt deep in my heart that this orb of lightning had come for me to take me to heaven. It felt as if I was communicating with it. I wanted more than anything to join together with its magnificence. I quickly plotted its trajectory and figured it would come over a certain house and then cross the street before going over the house on the other side. I felt this glorious moment was my destiny, so I started running as fast as I could to intersect the ball of lightning that I intended to jump into and ride into heaven.

My timing was correct, and the fiery chariot and myself were on a collision course in a few seconds. I was so happy to finally exit this world of suffering, lies, and pain. All I could think of was joining with this energy that reminded me of the glowing energy I often witnessed in the human body. The closer I got, the more I realized just how huge this glowing, whistling, oscillating ball of fire really was. It terrified and thrilled me to no end. I felt that I should simply jump out of my physical body and into this spiritual substance. I planned and counted my last few strides towards the lightning and pretended I was going to do a basketball lay-up into the sphere. I took the last step and leapt as high as I could towards the moving target. As I was coming into close contact with the fire-ball, I noticed it didn't seem to be hot, substantial, or aware that I was trying to jump into it. Then, an arm and a hand abruptly came thrusting out of the ball and struck my chest knocking me back, making me to fall backwards to the ground. The ball just kept on roiling and rolling along without touching the ground and ignoring my attempt to meld with it. I laid rejected on the pavement, stunned.

There were no words for what went through my mind at that moment. I was disappointed, totally bummed-out for missing my chance, befuddled by the experience, blown away by the arm and hand that I thought must have been God stopping my foolishness, and the horror that I would have to remain in this vale of tears. I slowly got up and looked at the ball continuing along on the same straight path skimming over house after house. Then, I remembered that ball lightning often goes out from a cloud and then returns on the same path. It was going to come back. I had another chance to try again. But I was confused, what if that was God's hand that stopped me the first time. I didn't care, I wanted to ride that fiery chariot to heaven even if it meant that I would spontaneously combust in a ball of flames. I had to know what would happen. So, I watched the ball return and planned to try again, even harder this time. But alas, when the ball lightning returned, I could not move at all. I was frozen throughout my body, it wouldn't move, and my feet felt like they were magnetized to the Earth. Much to my chagrin, the chariot of fire passed me by without noticing me at all and returned along its original path back into the clouds and disappeared, taking the rest of the thunderstorm with it into the sky above.

I stood watching for a long time in dismay, wonder, and disappointment. Then, above the clouds, other colorful light phenomena began to shoot red lightning high into the sky above, as if the ball lightning had risen into the upper atmosphere leaving fingers of red lightning streaking behind it. After a while, the clouds all settled down as the storm passed. Needless to say, I was never the same afterwards and my insatiable desire to understand what I had witnessed never left me.

The inner wisdom of plasma is illuminated by many stories like the one above. Saints have been "visited" by lightning or ball lightning that seemed to have inspired them for the rest of their lives. Anyone that I have ever met that has encountered ball lightning has told me that they were never the same afterwards. Even witnessing plasma in the aurora borealis changes people's lives. When people are in the arctic watching the cloud-like oval of the aurora above their heads, there is usually not a dry eye among them. Full-grown men weep when they directly experience plasma in the aurora. I have watched this emotional reaction happen to people at lower latitudes also when the aurora borealis is dramatic. Directly witnessing plasma in nature is a powerful and life-changing event. This experience is similar to when people, who have trained themselves to see the plasma flows in the human body, first start seeing the human aura and the petals in the chakras. It is shocking. It is like seeing the power of spirit manifest. Spirit which is usually invisible; we are coming into some type of new supersensible perception to witness plasma and the ethers.

Lightning is Plasma

Rudolf Steiner tells us that primal plasma in space help create life and humanity, and is a world of ideas. He even says that, 'plasma and flesh: they are basically the same substances.The human germ cell itself is eternal and only half of the germ plasma goes over into the new body.

Plasma works with all seven ethers and can be found when light ether "thrust down" to create electricity (static) energy in subnature; sound/chemical ether thrusts down to create electromagnetism, and life ether (the Harmony of the Spheres) is thrust down to create a "terrible force" [the "Third Force"]

The atmospheres of Earth are a living cycle of oxygen, lightning, and water. Water is 'fertilized by warmth and light from the universe.' Lightning (plasma) is all around us throughout the summer, fertilizing plants in a "flash of light." Everywhere in summer "lower lightning" (plasma) is all around us and lightning happens in us all the time because our thoughts are 'lightning and thunder (plasma) flashing within us.'

Steiner tells us that lightning (plasma) is like blood, and thunder is like thought and the nervous system. He says that 'God thinks in rolling fire.' Lightning (plasma) breaks through into the sensory world radiating in from the spiritual world. Stars (plasmas) appear to us out of the spirit like 'living revelations of lightning' (plasma). Behind the beings who live in lightning flashes are the Seraphim, Cherubim and Thrones. The Cherubim show us their tools in lightning (plasma). Lightning is the tool of the Cherubim, the fiery swords of the Cherubim.

According to Dr. Steiner, lightning derives from "universal heat" or "cosmic heat" which is the speech of the Seraphim who reveal themselves by shining. In thunder we hear the Thrones echo lightning, the speech of the Seraphim.

Thought originates in our soul like thunder rolling in the cosmos. When lightning flashes in the air, and when air and fire interact to produce thunder, this corresponds to the fire of our blood and the activity of our nervous system. Lightning in clouds corresponds to the fire and warmth in our blood, whereas the air outside, including all the elements it contains in the universe, corresponds to everything that passes through our nervous system. And just as lightning in its counterplay with the elements produces thunder, so the counterplay of blood and nerves produces the thought that flashes up in the soul. That is the invisible God who weaves and radiates through the universe. Steiner informs us that 'His warmth is in the lightning, His nerves in the air, His thoughts in the rolling thunder; and it was He who spoke to

Moses in the burning bramblebush and in the lightning fire of Sinai.'

Lightning comes out of the overheated air, the overheated universe, the overheated cosmic gas (plasma). As a consequence of the inner overheating of the cosmic gas (plasma), the warmth living in the cosmic gas (plasma) comes out. As a consequence of the greatly overheated cosmic gas (plasma), lightning flashes out. Lightning is not caused by electricity, but by the fact that the air is getting rid of, emptying away, its own heat. Steiner frankly tells us that: "It is lightning that stimulates electricity; lightning itself is not electricity."

The Cherubim are the directing powers in cloud formation, and the Seraphim are the directing powers in the lightning flash that issues from the clouds. If we look upon the expulsion from Paradise as really, referring to a descent from the periphery, we are describing almost word for word how man fell through his own weight, and how he had to leave behind him the forces and the Beings who form the clouds and the lightning—the Cherubim with the flaming sword. Man falls from the Earth's periphery, out of the region where the Cherubim hold sway with their fiery swords of lightning. The ancient seers who gave us *Genesis* gazed with full powers of seership into the life of humanity weaving in the etheric heights, before he fell from the regions where the Seraphim and the Cherubim hold sway.

Outer Knowledge, Inner Wisdom, Secret Teachings

The secret teachings of plasma were given to only a few because the power of plasma holds the power of life and death and cannot be given into the hands of immoral people. Dr. Rudolf Steiner could witness plasma and the etheric formative forces, which are usually invisible, through his inherent natural clairvoyance. He taught these secrets to only one person that we know of, Guenther Wachsmuth, his personal secretary. Wachsmuth wrote the book, *The Etheric Formative Forces in Cosmos, Earth, and Man* as the only exposition on ethers and plasma approved by Rudolf Steiner. Plasma, in Steiner's day, had just been discovered (1879) and he didn't want to give out the secrets just yet, even though, in the Wachsmuth's book can be found the secrets of plasma that are only now being discovered by scientism. That is why we condensed Wachsmuth's book, extracted the principles, and presented the content with commentary to help point out the secrets hidden therein.

There are many primal secrets and future technologies that will come from studying plasma. Scientism is at a loss to define plasma, or for that matter electricity, electromagnetism, nuclear energy, or the ethers that create all life on Earth and in the cosmos. We have written the book, *The Eternal Ethers: A Theory of Everything* to elucidate the nature, functions, and workings of the seven ethers but also felt it necessary to show that the ethers work together with plasma. Ethers and plasma are opposite poles of creation. Plasma is "not mater" and neither are the ethers. There are many similarities because plasma comes into existence using combinations of the ethers, based upon whether you are examining the cosmos, Earth, or humanity.

Atomic-based materialism cannot fit a "non-spatial" energy or force into their incomplete cosmology based upon the fundamental forces of gravity and atoms. Scientism is not "settled science", because the theories change constantly. For instance, you might wonder why scientism does not teach people about what they think plasma is and its critical role in life. This is because, if plasma does what scientism believes they observe, then everything scientism has ever taught is null and void, useless, and meaningless data without cohesion or comprehensive application. Scientism has enslaved electricity into the marvels of the technological age and yet cannot tell anyone where electricity comes from, how it does what it does, or explain what it is, comprehensively. Scientism is a series of changing estimations and best guesses that miss the mark when creating weapons or experimental medical treatments for humans without first thinking about the health and safety of humans. Scientism has become a new God-less religion that ignores more than they know and claims to understand things that they don't, denying the wisdom and secret teachings of the past. Scientism works in a vacuum with one-eyed color-blindness, half-deaf perception, and no consideration for the invisible forces and beings in nature all around us. Thus, the secret teachings of plasma and the seven ethers has remained a secret to this very day as scientism continues to torture and weaponize what they think plasma is in their laboratories.

What is plasma? What will we find once the wisdom and secrets of ethers and plasmas are revealed? Plasma is the Tree of Life, a name given by Steiner to chemical/sound ether and life ether working together with combinations of the warmth ether and light ether to create the entire universe, Earth, and the human being's physical and etheric bodies.

To answer the mysterious question: What is plasma? It is almost impossible to do so without understanding the complete nature of the seven ethers. Wachsmuth's book on the ethers is the most complete expose on the ethers that had Rudolf Steiner's approval. Before we endeavor to understand the ethers, we must undo and unwind some of scientism's theories that roll around in the heads of modern materialists. First, scientifically speaking, the cosmos is nothing but the fundamental forces of gravity, the strong nuclear force, and the weak nuclear force. Unfortunately, once you understand that atoms are theoretical and only exist in Earth's immediate environment and 1% of the universe, then the two nuclear forces disappear.

Once you understand the secret teachings of plasma and ethers you will realize that gravity is another "made up" force that only exists in theories. Scientism's math can't explain the 99.9% of the plasma universe with only atoms and gravity which only work comprehensively in Earth's atmospheres. Science theories stop being effective when we enter the exosphere of the Earth and certainly cannot explain the plasmasphere, the Sun/Earth magnetic reconnection, or the solar wind and solar forcing of electromagnetism, plasma, CMEs, proton winds, or any of the mechanisms that create life on Earth. Science does not have a comprehensive theory of life.

Let us take a simple example of how water and oxygen come into Earth's atmospheres. Scientism theorized for a long time that all water and oxygen on Earth came from asteroids carrying them down from space where elemental (ionized) water and oxygen do exist. Any child could figure out why that is nonsense. Eventually, scientific materialism decided on a new theory which goes something like: We don't fully understand where water or oxygen come from. Ultimately, there is no generally accepted theory about the creation of the two most important elements necessary for life on Earth. Some researchers now say that: We "believe" that water and oxygen both came from ionized elemental particles in space (the "vacuum" of space), called dust, that somehow organized into molecules of water and oxygen and somehow came into Earth's atmosphere. There you have it. Scientism has a clue, but not enough wisdom and secret teachings to understand that the substances of life comes from what some traditions call the "Tree of Life"; from which, flows the Waters of Life and its leaves are "for the healing of the nations" and its fruit gives immortality to those who can rise up to New Jerusalem to receive these divines gifts of food and drink. Now this may sound silly to scientism, but it is an accurate analogy of what Rudolf Steiner gave as the secret teachings concerning the phenomena of life and the forces and beings that create life on Earth.

Spiritual scientific research, which we have presented above, demonstrates that ionized helium is transformed by plasma through the rays of the Sun that contain warmth, light, chemical/sound, and life ether, transforming elemental helium into ozone (O3) in the ionosphere, which protects humans from the harmful rays of the Sun. This "ozone layer" is invisible plasma that protects us and makes oxygen from ozone in the lower atmospheres. Plasmoids in the thermosphere (6,000 degree +) take the ionized helium particles of "dust" and heat them to temperatures of one million degrees or hotter in complex dusty plasmoids to create what scientism calls "fusion" energy, the same energy they imagine drives the Sun's atmospheres.

The Sun's solar wind (breath) is constantly recharging the plasmasphere, which is held together and flows along magnetic field lines called Van Allen belts. This cold plasma from the solar breath enters the plasmasphere and is recharged by a huge magnetic rope (plasma filament) every eight minutes from the Sun through magnetic transfer flux portals in the Earth's outer magnetosphere. In this Sun/Earth symbiotic process, a magnetic reconnection portal opens that allows an Earth-wide tube of fresh plasma from the Sun to pour into the plasmasphere expanding or contracting the parameters of the plasmasphere like human lungs breathing. This fresh breath of plasma from the Sun re-enlivens the plasmasphere with cold plasma that has not yet been aligned with Earth's magnetic fields, the magnetosphere, and the plasmasphere. The electromagnetic plasma flowing through these magnetic transfer flux points are portals to the Sun that reconnect every eight minutes to renew magnetic reconnection that ultimately pulls the Earth behind the Sun's movement through space and feeds the Earth fresh plasma. This reconnection is instantaneous and happens immediately over the 93 million miles between the Sun and Earth.

Plasma in the plasmasphere is cold and does not generally form into plasmoids until the auroral electro jet moves cold plasma from the plasmasphere into the exosphere and thermosphere, thereby creating the plasma sheets of the norther lights, the aurora borealis, which meet ionized elements of carbon, nitrogen, oxygen, hydrogen, and phosphorus gases (plasma) rising up from the Earth towards the Earth's north pole. When plasma flows into the north pole it aligns cold plasma with the Earth's many magnetic forces falling in through the north pole to the Earth. This movement of cold plasma into Earth's magnetic fields drives plasma flow down into the thermosphere where complex dusty plasmoids absorb cosmic and terrestrial "dust" particles and spin them into electromagnetic filaments, cables, ropes, and strings that eventually create a torus shape wrapped with dual-strands of electrical charges. These double-layered plasmoid balls are magnetohydrodynamic fusion generators creating primal life forces, both organic and inorganic.

In complex dusty plasmoid fusion cells, primal helium and hydrogen start the process of organized condensation with ozone and hydrogen peroxide, which is abundantly available in space and the Earth's exosphere and send their "transmuted" particles lower into the atmosphere where they become primal water and oxygen. This process is continuous and was going on long before humans descended into the material/physical realms. In ancient times, this process was seen as a great tree (Yggdrasil in North Myths) growing into the heavens. All nine realms of gods, nature spirits, humans, ice giants, dragons, and other beings lived in different realms amongst the roots, trunk, and limbs of the World Tree. A dragon wound around the tree just as the serpent wound around the Tree of Knowledge.

In the living imagination of a World Tree that feeds and nourishes humanity, the trunk is the axis of the Earth, the roots are the south pole, and the upper branches reach up through the north pole into the exosphere and the plasmasphere where the leaves of the tree fill the lungs of the plasmasphere with the Breath of God from the Sun. The thermosphere filled with plasmoids is like the fruit of the Tree of Life, and from the dew of the tree comes the Waters of Life bringing the "waters from above" to the lithosphere. These "plasma secrets" are implicit in Steiner's teachings concerning the ethers, as you probably noticed in the previous chapters derived from Wachsmuth's book.

Still the question lingers: What is plasma? Since we have discarded atoms and the strong and weak nuclear force by showing that atoms didn't just "come into existence one day"—since the universe is less than 1% atomic substance with an imaginary nucleus and little electrons spinning in clouds around them. Atoms, or monads, came into existence by the forces in plasmas and plasmoids working with the ethers from the Sun to structure, build, and create electromagnetically stable molecules in the thermosphere. Atoms don't exist. Dr. Steiner tells us that the atom is really a "monad" that replicates the electromagnetic, plasmic, and etheric mechanisms of the "life body" of the Earth as a whole. In other words, the Earth is one big primary particle, monad, or "atom" (in scientism) and every cells and monad in the human being replicates those same mechanisms. Which begs the question: Are plasmas and ethers alive or living sentient beings, and do they live inside of me?

Plasmas and ethers are impossible to describe with terms and thoughts taken from the material world because they are non-spatial, filled with etheric geometry, seemingly universal and eternal, able to contain fusion for creating molecules, masters of levity and gravity, able to control transmutation without an "explosion", monad incubators, the waters of life, invisible, and generally magical non-substance. Magic is the word humans have always used to describe what they could not understand. Plasmas and the ethers are simply magic, in a word.

Scientism tells us false fairy tales that lack imagination, creativity, or the truth of reality. Let's take the first children's creation fairy tale called the Big Bang Theory. This unimaginative tale had to have the concept of "Inflation" added to the theory to try and explain all the gross incongruencies that facts presented to observers through their telescopes and devices. Recently, scientism had to change the "settled science" theory of the Big Bang to add three new stages beyond "Inflation", called Stars III, Stars II, and Stars I. While the Big Bang was happening, when that small ball of proto-matter was exploding into the vast organized and multifarious universe, some early stars in phase one (Stars III) came into existence that were only hydrogen and helium proto-particles that "somehow" turned into the entire amount of water we find in the universe. Hmm. Then, stars in phase two (Stars II) came into existence (magically) and created heat-light, even though gravity and atoms had not come into existence yet. Hmm. Then, phase three happened and new stars magically came into existence (Stars I) which created matter, as we know it now, and gravity that ensued and then ultimately controlled the universe. This is, of course, a different story of creation than the Big Bang first told to us, and not a single scientist has come forward to say they were wrong, or lied, or have kept lying ever since and made sure the textbooks don't indicate past mistakes and bad theories.

Dr. Rudolf Steiner's cosmology points out that the universe was created by the actions of spiritual beings who donated substances in the earlier "incarnations" of the Earth. We might call them Earth I, Earth II, Earth III, and now we are in Earth IV where matter surrounds us. In Earth I, the warmth ether started moving like a nebula of invisible dust (plasma) until warmth was created in clustered balls of energy. This is much like the plasmasphere where primal substance is not organized and swirling around as groups of cells. Thus, Stars III should be Stars I; they got the right substances but the wrong order. Earth II worked through the forces of light ether to create the primal oxygen/air. Earth III used chemical/sound ether to create a fluid environment. Earth IV used life ether to add the human being to a world of minerals and material substances. Thus, warmth, light, chemical/sound, and life ether have manifested the substances that eventually evolved into (congealed into) our earth, water, air, and fire. Thus, the four stages of the Big Bang are also found in Steiner's "Incarnations of the Earth" and create the same "substances" of primal life: minerals (life ether), water (chemical ether), air (light ether), and fire (warmth ether). These later evolved on Earth IV into minerals, plants, animals, and human life.

One might think that we are saying that an understanding of plasmas can only arise once an understanding of the seven ethers is obtained. This is true; or, you can read all that we presented in the first third of this book (outer knowledge) and then see if you can put together a complete cosmology of plasma and its workings. This task is doomed to fail because science does not consider that which it cannot sense with the limited five senses, and plasmas and ethers are often not invisible or perceptible to limited human senses. Scientism works with lab-made plasmas but have not been able to control or understand it. Spiritual science, on the other hand, prepares one to witness the four ethers that manifest on Earth, which is a good preparation for envisioning the scientific laws that plasma "breaks." We have to unwind and unlearn those old, worn-out theories of science to open our minds to cognizing plasma "substance" that is not matter, in plasma fields that are invisible, making plasmoids in etheric realms that human minds can't conceive of, and doing things materialists would call impossible. And that leap of cognitive dissonance is followed by imagining the plasma wonders of quasars, black holes, a million light-year long plasma filaments connecting galaxies, plasma clouds filled with self-organized cells that have all the characteristics of life, as well as the well-hidden plasma in the human body. Plasma and the ethers open the mind to the forces and beings of life and the primal creative spiritual beings of the cosmos.

In many of our books, especially *The Eternal Ethers*, we reveal the magic of the fours ethers, the unfallen ethers (Tree of Life), and the akasha ether (Mother of All). We show that ancient traditions have known for millennia the mysteries of ethers and plasma, both in the human "electric body" and the electromagnetic actions and creations in the cosmos. The Earthly and Cosmic Nutrition Stream, also called the Etherization of the Blood, is explained in great detail through the workings of the etheric formation forces and also through the elements and physical mechanisms that drive the continuous feeding of the gods and their reciprocal feeding of humans. Suffice it to say, the same processes of the Earth are duplicated in a smaller version in the human being. For example, the Etherization of the Blood is very similar to plasma flowing into the magnetic north pole creating the aurora borealis. From heart to head, ether streams carry invisible plasma (electromagnetic energy as combustion of ionized elemental substance) up to the pineal gland where an internal aurora is created around the corpora quadragemini and pineal gland, stimulating piezoelectric discharges from the calcium carbonate crystals in the pineal body. When discharges from the pineal gland enter the cerebral spinal fluid in the fourth ventricle, which is surrounded by the strongest electromagnetic field in the brain located in the cerebellum, the discharge moves through the third ventricle's fluid that is filled with minerals and metal particles descending from the first and second ventricles into the cerebral spinal fluid filling the third ventricle. This piezoelectric discharge causes the third ventricle to "light up" with an electric charge that stimulates the pituitary gland causing it to secrete pituitrin into the blood and nerve systems of the body. This process is a magnetohydrodynamic generator that creates the electromagnetic field emitted by the head.

We can see in the comparison of the human head and the Earth's north pole that the action of plasma and the ethers working together create a display of light and color in the aurora and a white "halo" around the north pole is similar to the plasma rising in the etheric stream from heart to head that creates a rainbow around the pineal gland and an oval glowing light around the head of a spiritual person that a clairvoyant can see. This is described in great detail by many spiritual traditions that have secret teachings concerning these forces and the beings who create them. We can also see in the Tree of Life analogy the plasmodic creation of water and oxygen which are the processes of spiritual nutrition that can be found in the ancient traditions concerning spiritual nutrition. The clearest traditions are those that concern the human heart.

Cosmically speaking, plasma is gravity, and gravity is life ether which can act in a cosmic or earthly manner sucking matter to its being, holding "matter" together. Gravity creates the pulse beat of the contracting heart, which drives the motion of the heart as the heart "falls" into the suction force of Earth's gravity, or life ether. The human heart, which is the biggest dynamo of bio-electromagnetism in the human body, is another magnetohydrodynamic generator that produces a large magnetic field that extends out about 10-12 feet from the human body. The plasma generator in the head, that we described above, is at least 50 times weaker than the heart.

The third magnetohydrodynamic (MHD) generator in the human body has now been found in the Enteric Nervous System (ENS) in the lining of the intestines where the longest peripheral nerve in the body is found. The ENS is called the "second brain" which can create coherence with the heart and brain generators via the vagus nerve, the longest nerve in the body (10th cranial nerve). These three generators must be synchronized and coherent for the human being to bring into balance their capacities of thinking, feeling, and willing. All three generators are plasma generators that create bio-electromagnetism that feeds the nerves, blood, tissue, and all aspects of the human body that combine to create a platform for individualized self-awareness and the ability to consciously reflect upon one's thinking, feeling, and willing. All three generators can be explained by scientific evidence, facts, data, and measurements, but they still cannot be seen, explained, or enhanced by scientism. Many spiritual paths describe these mechanisms in marvelous detail and explain what these forces are, and WHO they are. What they all do agree upon is that these cosmic "forces" are not the "Beings" behind them. Some type of invisible, cosmic Being donates and controls those forces with "intelligent design" that looks a great deal like life itself. In other words, spiritual beings work through the forces of ethers and plasmas, but they are not those forces—they are emanating those forces, like a person in the three-dimensional world can cast a shadow into a two-dimensional world. There are big differences between Beings and forces.

If ethers and plasmas are the forces that came from Beings, then who are the Beings behind these forces that mold and shape matter on Earth and in the cosmos? Obviously, if the forces are invisible, but we can work with and manipulate those forces, then the Beings themselves are invisible and almost unimaginable to child-like humanity. If we are to believe in scientism, these forces have been around for about four or five billion years slowly building the world we see with our limited senses. Humans are in a nursery being incubated like the cosmic particles of helium and hydrogen in space are birthed into inorganic life in the spheres around the Earth. Beings work behind and through plasmas and ethers as they descend into matter, like ice crystalizing by the force of cold. Matter is precipitated over four stages of condensation into the kingdoms of nature: mineral, plant, animal, and human. Humans are like the fruit of the Tree of Life dropping from the Garden of Eden into the world of cold, hard matter. Once the fruit of the Tree of Knowledge of Good and Evil was picked, the garden disappeared along with the Tree of Life that elevated itself to a place where humans cannot pick its fruit so easily.

The Tree of Life is still in the Garden of Eden on the white island that the aurora borealis makes above the north pole. A conscious spiritual aspirant can cross the rainbow bridge into that realm once the seeker understands the other eight realms in the World Tree and the process of the Etherization of the Blood, the Earthly and Cosmic Nutrition Stream.

The realm of the gods at the top of the World Tree was called Asgard where the Norse gods and goddesses lived, far above the land of humans called Midgard. Asgard can be imagined as the realm of unfallen chemical/sound and life ether that all humans pass through when they go to sleep and pass into the outer realms of the planets. Most people are unconscious during sleep when they pass through this realm. Initiates are awake in their sleep and understand that they go into these etheric/plasmic realms at night like a student going to school or a hungry person going to eat. The rotating white aurora is the portal to spiritual realms that are far beyond what we have described so far. This is the realm called Shambhala, Mount Meru, Tushita Heaven, New Jerusalem, the White Island, the Temple of Wisdom, the Hall of Wisdom, and many other names. It is the realm of plasma/ether which is timeless, non-spatial, filled with consciousness and archetypes, and defies all Earth-bound knowledge based on the material world.

After all of these unimaginable ideas have been crudely presented with Earth-bound words, we need to now consider the genesis and other aspects of the ethers and plasmas. The primal ether was the Mother of all ethers called akasa, akasha, Aviditi, Vac, Fohat, Solar Magnetism, the Mother Lodge of the Sun, the Solar Hero, and many other names. It is the unmanifest that is manifest, the Alpha and Omega, the original ground of being, spaceless space or timeless time with a consciousness that was born before consciousness. Many call akasha ether, the Akashic Records where everything that ever existed or happened is recorded and can be accessed by moral initiates.

Akasha was the first mover, the original breath, the magic word OM, the Word, the Logos, and the original creator. Akasha, according to the ancient stories, gave birth to warmth, light, sound, and life in their original nature. Akasha is all around us but is nowhere. There is no explanation that can describe akasha. It is the non-vessel of primal-space that held creation and started the original spiral movement of the ocean of primal material.

Just as mystical and sacred to the ancients were the unfallen ethers of chemical/sound and life ether. These two perfect ethers from the Sun were seen as life itself. Usually, the ancients only saw the power of the Sun in chemical/sound and life ether when there was a cataclysm with plasma on the Sun. They understood these two forces to be the greatest powers in the cosmos. Fortunately, solar CMEs and solar cataclysms happen infrequently, but they were not forgotten as the force of the chemical/sound and life ether bombarded the Earth. The ancients respected the Sun and the beings who live there and knew that a Solar God watched over the Earth to make sure the power of the Sun would only be gradually absorbed into the human being. This loving Solar God was found at the teachings of every culture and religion. Humans inherently knew that the Sun, who created us, would not abandon us. Therefore, to connect to the life-giving ethers and plasmas of the Sun, the solar deity, Christ, descended to Earth and connected the Sun to the Earth for all time to come; not unlike the Sun/Earth magnetic reconnection that happens every eight minutes or so and has birthed life on Earth.

It was this teaching concerning the Solar System that was secret in ancient times. Before Jesus of Nazareth became the vehicle for this Solar Deity, all religions had teachings about the descent of a god from the Sun, in one form or another. Probably the best pre-Christian personification of the Love of the Sun was the ancient Indian belief in Fohat, a solar magnetic force of fire that reaches into the heart of humans and sets the spark of spirit on fire. This being called Fohat had been around since the creation of the cosmos and the idea of the human being was planned and created in the Mother Lodge of the Sun by Christ, the Second Person in the Holy Trinity, and the guiding being of the Second Rank of spiritual hierarchical beings called: Spirits of Wisdom, Spirits of Motions, and Spirits of Form. Fohat was a personal god who descended on the rays of the Sun and found a home in the human heart. Fohat is inexplicable, inexpressible, unnamable, cannot be contained, and moves through all worlds like a blazing fire.

Thus, we can see that no book can contain the full description of the types of plasmas and how they function. Even a comprehensive understanding of the ethers, both fallen and unfallen, only prepares you to understand plasmic electricity and the mechanisms by which it helps the etheric formative forces bring life to birth and sustain it with earthly and cosmic nutrition. We hope that our efforts here will spark a flame of unquenchable desire to study diligently to understand the forces and Beings who have worked so hard to provide a realm where individualized human consciousness can come to birth and find freedom and love everywhere. As Rudolf Steiner said: the cosmos is filled completely with an unimaginable amount of love.

Appendix

Recent Solar Emissions Decrease Ozone in the Upper Atmosphere

Ozone Decrease Observed in the Upper Atmosphere May 11, 2024, Mother's Day Solar Storm

Alexandre Winant, Viviane Pierrard, and Edith Botek
From: Annales Geophysicae, European Geosciences Union

https://angeo.copernicus.org/preprints/angeo-2024-29/

Abstract

On May 11th 2024, a succession of coronal mass ejections that merged together struck the Earth and induced large scale perturbations in the magnetosphere. During this event, satellite observations showed a large solar energetic proton (SEP) event associated with an extreme geomagnetic storm. At the same time, satellite observations of atmospheric ozone have been performed by AURA/MLS. In this work, we present the first observations of the effect of the storm of May and the following SEP of June 8th on ozone concentration throughout the atmosphere. Observations of the MLS show that the event of May lead to stronger depletion of O3 in the upper part of the atmosphere than in June. This difference is explained by the type of particle precipitation that occurred during the two events, with both protons and electrons in May and only protons in June. Neither event caused ozone depletion in the stratosphere while strong decreases are observed in the mesosphere. In May, mesospheric ozone depletion is observed during 18 days and reaches a maximum of 60 %. In addition, the storm of May also caused a noticeable decrease in ozone concentration (up to 20 %) at altitudes above 90 km.

Teilhard De Chardin's Spiritual View of the Ether

Piere Teilhard de Chardin's war-time letters to his cousin Marguerite Teilhard-Chambon, (*Writings in Time of War*, Pierre Teilhard, Translated by Rene Hague, Harper & Row, New York, 1817), will have familiarized those who have read them with the early stages of his thought, and drawn their attention to the existence of a number of essays written at the front during the same period.

> "He asked himself whether he would ever be able to disclose all that he had seen and felt and thought during this amazing period of transformation. Just before his demobilization, he wondered anxiously. Will they ever listen to me?"

It is quite staggering to see a man working as a stretcher-bearer in that ghastly fighting, living for the most part in the mud of the trenches, and at the same time taking advantage of his too brief rest periods to jot down on paper his notes and plans to be worked up later into essays that dealt with the most profound problems.

As each was finished, it was sent off to his cousin or to his sister Guiguite, or to one of his fellow Jesuits, sometimes in the hope that it might be published. At times he would ask his sister or his cousin to have some copies typed for his friends. After a first draft, in which there were a considerable number of alterations and deletions, he would generally make a fair copy. This explains how it is that we have two copies of some of the papers, with, of course, a number of variations. This is the case with *Mon Univers,* 1918.

Shortly before her death Marguerite Teillard-Chambon bequeathed these manuscripts to her sister Alice, who agreed to their publication. It must be remembered, too, that he was writing primarily for himself. He was drawing up a testimony, almost, as we have seen, a testament. He communicated his work to a restricted circle, with no hope of having it published. He had tried to do so with two articles on current problems; one was rejected, and the other was accepted after cutting.

Teilhard's *Cosmic Life*

Cosmic Life was finished on 24 April 1916, at Nieuport. It is the first of Teilhard's extant writings to be fully characteristic of its author. Knowing what risks he was exposed to at the war's front, Pere Teilhard de Chardin wrote it as his 'intellectual testament', and it contains in embryo all that was later to be developed in his thought: in embryo, because what he has to say is less clearly focused, less accurately expressed and less completely proved than in later writings. Some points, indeed, that he was later at pains to clarify, are presented in this essay simply as the evidence of Revelation. As in a number of other essays in this volume, the style in which the ideas are expressed is lyrical and elevated. In Pere Teilhard's own words, it is the fire in his vision that he is trying to communicate.

Extracts from *Cosmic Life*

...The first aspect of this profound oneness of all the elements of the universe is their common 'rooting' in the mysterious and pre-eminently cosmic entity that we call the *ether*.

In spite of the strange properties that make it as physically real as a block of stone and at the same time as intangible as an abstract limit, physics is inexorably bound to accept the *ether*. It is the medium that must be posited for the transmission, or even perhaps the dispensing, of transient energies, to maintain or even to hold in tension the links that attract or repel the particles into which the world breaks down. Moreover, it is the ultimate term at which the cosmic particles are resolved whether we regard them as eddies produced within a primitive homogeneous fluidity or simply as the countless centers around which one and the same fundamental substance radiates and folds itself. In either case, whether as the primordial stuff of things or as being the universal active medium (like hyle or energeia) it enters into our view of the world, by reason of being by nature the ultimate support of substance and activities, as a reality whose uniform plenitude can admit neither hiatus nor break, either of those would open the door to non-being. *The ether is like some great reservoir of elastic fluid that could be twisted again and again without the least rift or break appearing*, to isolate from the total mass the individual local realities produced by the twistings, however complex and independent of one another we may conceive them to be. At the same time, it is a fluid whose nature is such that we can have no empirical knowledge of it: no natural constitutive part can be distinguished in it, nothing that resembles an atom or a molecule. *Force cannot break the continuity of the ether, nor can it be broken down by analysis*. We should think of it not as having the texture of a liquid into which the countless hordes of a thousand and one different particles are crowded together, but as the unimaginable state of some center that is indefinitely extended in space, everywhere the same and yet everywhere different from itself.

...Everything in the universe, we must realize, proceeds from the *ether*, or in the *ether*.

...I can put my hand at random on any element in the teeming multitude and if I confront it with another one, also chosen at random, I am obliged to admit that in both a real inner identity lies hidden — that identity may, no doubt, be completed by an individual and incommunicable immanence, but it is not destroyed by it. It is identity in the *ether*, whose unique center, dispersed through all things, is the *prime matter* — indivisible in spite of its immensity — of *all that springs up within the vast cosmos*. Like knots spaced out along a cord, or like the folds into which a single curtain falls, or the eddies forming on one and the same surface, everything that moves and lives in the universe represents, in one particular aspect, the modifications of one and the same thing; and every monad, if it looks into itself, can find that thing as the initial point at which all things make contact in their inmost essence.

...there are without any doubt large over-all currents that affect the multiplicity of atoms and stars, and through diem the common soul of an evolution is infused into the common body of all that has its basis in ether.

If we look at this world, we see that the fundamental substance within which souls are formed, the higher environment in which they evolve — what one might call their own particular Ether — is the Godhead, at once transcendent and immanent, in whom we live and move and have our being.

God is vibrant in the ether; and through the ether he makes his way into the very marrow of my material substance. Through Him, all bodies come together, exert influence upon one another and sustain one another in the unity of the all-embracing sphere, the confines of whose surface outrun our imagination.

God is at work within life. He helps it, raises it up, gives it the impulse that drives it along, the appetite that attracts it, the growth that transforms it. I can feel God, touch Him, live Him in the deep biological current that runs through my sod and carries it with it.

I felt my body, my soul, and even my spirit pass into the ethereal tint, unreal in its freshness, that caressed my eyes. Serene and iridescent, its color bathed more than my senses; it in some way impregnated my affections and thoughts. I melted away in it, lost in a strange yearning to attain some individuality vaster and simpler than mine — as though I had become pure light.

With pure and generous love, a new energy penetrated into me—or emerged from me, which, I cannot say—that made me feel that I was as vast and as loaded with richness as the universe.

First there is the revelation of the unique matter, and then, even more wonderful, that of the unique soul: first, what the laboratory shows us, and then what nature does; first the ether, and then life. It is not, you see, enough to hear what science has to tell us, and to observe from outside the cosmic currents taking shape, in their individual eddyings or their over-all drifts; it is those currents that make us what we are, it is through us that they run, and we must be able to feel them.

We can see, then, that what characterizes the Universal Element as we find it realized in Christ, is not that it is a quasi-matter, a plastic or informable element, an agent of absorption, but a quasi-soul, a plasmatic or informing element, a force of determination.

Recommended Reading

Gabriel, Tyla. *The Gospel of Sophia: The Biographies of the Divine Feminine Trinity,* Volume Our Spirit, Northville, 2014.

Gabriel, Tyla. *The Gospel of Sophia: A Modern Path of Initiation,* Volume 2. Our Spirit, Northville, 2015.

Gabriel, Tyla and Douglas. *The Gospel of Sophia: Sophia Christos Initiation,* Volume 3. Our Spirit, Northville, 2016.

Gabriel, Douglas. *The Spirit of Childhood.* Trinosophia Press, Berkley, 1993.

Gabriel, Douglas. *Goddess Meditations.* Trinosophia Press, Berkley, 1994.

Gabriel, Douglas. *The Eternal Curriculum for Wisdom Children: Intuitive Learning and the Etheric Body.* Our Spirit, Northville, 2017.

Gabriel, Douglas. *The Eternal Ethers: A Theory of Everything.* Our Spirit, Northville, 2018.

Gabriel, Douglas. *Hidden History of the Grail Queens.* Our Spirit, Northville, 2019.

Gabriel, Douglas. *The Spirit of Childhood*: The Waldorf Curriculum. Our Spirit, Northville, 2024.

Gabriel, Douglas. *Intuitive Learning: Activating the Well-springs of Giftedness.* Our Spirit, Northville, 2023.

The Series: *From the Works of Dr. Rudolf Steiner*

Gabriel, Douglas. *The Incarnation of Ahriman: The Occult Annihilation of the Soul.* Our Spirit, Northville, 2023.

Gabriel, Douglas. *How to Become An Angel: Preparing for the Future Sixth and Seventh Epochs.* Our Spirit, Northville, 2023.

Gabriel, Douglas. *Surviving Apocalypse: Creating Your Ark.* Our Spirit, Northville, 2023.

Gabriel, Douglas. *Life After Death: Steiner's Book of the Dead.* Our Spirit, Northville, 2023.

Gabriel, Douglas. *The Second Coming of Christ in the Etheric Realm.* Our Spirit, Northville, 2024.

Gabriel, Douglas. *Grappling with Evil: Conquering Fear and Anxiety.* Our Spirit, Northville, 2024.

Gabriel, Douglas. *The Human Heart: Supersensible Organ of Perception.* Our Spirit, Northville, 2024.

Gabriel, Douglas. *Anthroposophia: The Divine Feminine Trinity.* Our Spirit, Northville, 2024.

Gabriel, Douglas. *Spiritual Enlightenment and Initiation.* Our Spirit, Northville, 2024.

Gabriel, Douglas. *Cosmic Intelligence and Spiritual Hierarchies.* Our Spirit, Northville, 2024.

Gabriel, Douglas. *Developing Christ in Our Threefold Ego.* Our Spirit, Northville, 2024.

Gabriel, Douglas. *The Eighth Sphere: Evil and Evolution.* Our Spirit, Northville, 2024.

Gabriel, Douglas. *Meditation for Enlightenment and Ascension.* Our Spirit, Northville, 2024.

Gabriel, Douglas. *Dweller on the Threshold: Encountering the Lord of Karma.* Our Spirit, Northville, 2024.

Gabriel, Douglas. *Rudolf Steiner and the Masters.* Our Spirit, Northville, 2024.

Gabriel, Douglas. *The Inklings and Star Wars.* Our Spirit, Northville, 2024.

Gabriel, Douglas. *Perception as Revelation of the Spirit.* Our Spirit, Northville, 2025.

Gabriel, Douglas. *The Etherization of the Blood.* Our Spirit, Northville, 2025.

Gabriel, Douglas. *The Perfected Bodies of Jesus Christ in the Realm of Spiritual Economy,.* Our Spirit, Northville, 2025.

Plasma Videos

These are good videos about plasma, bioplasma, and the science of ascension.

https://www.youtube.com/watch?v=Ttb4Zefwxr8

https://www.youtube.com/watch?v=2dzPsIS8d6M&t=0s

https://www.youtube.com/watch?v=skN9_FfWcyA&t=0s

From Thunderbolt Project and the Electric Universe

https://www.youtube.com/watch?v=5AUA7XS0TvA&list=PLwOAYhBuU3UfvhvcT1lZA6KbSdh0K2EpH&index=2

https://mail.google.com/mail/u/0/?shva=1#inbox/FMfcgzQZTCgLSTbNpzNJRqSXfHCqrZMP?projector=1

From: https://www.youtube.com/@Versadoco

INORGANIC LIFE: the Ball Lightning, Part 1/3—VERSADOCO
https://www.youtube.com/watch?v=JyUQbSRr1X0

INORGANIC LIFE: Intelligent Plasmas, Part 2/3—VERSADOCO
https://www.youtube.com/watch?v=P1IkXXwWhwI

INORGANIC LIFE: Alive Patterns, Part 3/3—VERSADOCO
https://www.youtube.com/watch?v=zqJRzORLaZM

Ancient Plasma Art

https://www.youtube.com/watch?v=P65SvhtRH5k

Bibliography

Blattman, George. *The Sun, the Ancient Mysteries, and a New Physics.* Floris Books, Anthroposophic Press, London, 1972.

Blavatsky, Helena Petrovna. *Isis Unveiled: A Master Key to the Mysteries of Ancient and Modern Science and Theology*, Quest Books, Chicago, 1997.

Blavatsky, Helena Petrovna. *The Key to Theosophy*, Create Space Independent Publishing, 2016.

Blavatsky, Helena Petrovna. *The Secret Doctrine: A Synthesis of Science, Religion and Philosophy*, TarcherPerigee, 2009.

Blavatsky, Helena Petrovna. *The Voice of Silence: Being Extracts from the Book of the Golden Precepts*, Theosophical University Press, 1992.

Bockemuhl, Jochen. *Toward a Phenomenology of the Etheric World: Investigations into the Life of Nature and Man.* Anthroposophic Press, Spring Valley, N. Y., 1977.

Falck-Ytter, Harald. *Aurora: The Northern Lights in Mythology, History, and Science.* Floris Books, Antrhoposophic Press, London, 1983.

Forward, William and Wolpert, Andrew. *The Image of Blood: The Golden Blade* No. 48. Floris Books, Edinburgh, 1995.

Franke, Sylvia. *The Tree of Life and the Holy Grail: Ancient and Modern Spiritual Paths and the Mystery of Rennes-le-Chateau.* Temple Lodge, London, 2007.

Gebser, Jean. *Anxiety, a Condition of Modern Man*, Paperback, 1962.

Gebser, Jean. *The Ever Present Origin*, Ohio University Press, 1991.

Hall, Manly P. *The Most Holy Trinosophia of the Comte De St. Germain.* The Philosophical Research Society, Inc., Los Angeles, 1983.

Harrison, C. G. *The Transcendental Universe.* Lindesfarne Press, New York, 1993.

Jocelyn, Beredene. *Citizens of the Cosmos: Life's Unfolding from Conception. Through Death to Rebirth.* Continuum, New York, 1981.

Katzenellenbogen, Adolf. *Allegories of the Virtues and Vices in Medieval Art.* Medieval Academy of America, Canada, 1989.

Konig, Karl. *Earth and Man.* Bio-Dynamic Literature, Wyoming, Rhode Island, 1982.

Landscheidt, Theodor. *Sun-Earth-Man a Mesh of Cosmic Oscillations: How Planets Regulate Solar Eruptions, Geomagnetic Storms, Conitions of Life, and Economic Cycles.* Urania Trust, London, 1989.

Laszlo, Ervin and Kingsley, Dennis L. *Dawn of the Akashic Age: New Consciousness, Quantum Resonance, and the Future of the World.* Inner Traditions, Rochester Vermont, 2013.

Laszlo, Ervin. *Quantum Shift in the Global Brain: How the New Scientific Reality Can Change Us and Our World.* Inner Traditions, Rochester Vermont, 2008.

Laszlo, Ervin. *Science and the Akashic Field: An Integral Theory of Everything.* Inner Traditions, Rochester Vermont, 2004.

Lehrs, Ernst. *Man or Matter.* Rudolf Steiner Press, Forrest Row, 2013.

Loeteman, Kees. *Gaia-Sophia: A Framework for Ecology: A New Approach to Evolution and Working with the Environment.* Floris Books, Edinburgh, 1991.

Milner, Dennis and Smart, Edward. *The Loom of Creation: A Study of the Purpose and the Forces that Weave the Patter of Existence.* Harper & Row, New York, 1976.

O'Neil, George and Gisela. *The Human Life.* Mercury Press, New York, 1990.

Pfeiffer, Ehrenfried. *The Heart Lectures.* Mercury Press, Spring Valley, New York. 1997.

Prasad, Rama. *Nature's Finer Forces & the Science of Breath.* The Theosophical Publishing Society, London (1890).

Prasad, Rama. *The Science of Breath & the Philosophy of the Tatwas.* The Theosophical Publishing Society, London (1890).

Prokofieff, Sergei O. *And the Earth Becomes a Sun: The Mystery of the Resurrection.* Wynstones Press, Stourbridge, England, 2014.

Prokofieff, Sergei O. *The Appearance of Christ in the Etheric: Spiritual-Scientific Aspects of the Second Coming.* Temple Lodge, London, 2012.

Ragon, Jean Marie. *Esoteric Studies in Masonry.* Daath Gnosis Publications, USA, 2015.

Spengler, Oswald. *The Decline of the West*, Vintage, 2006.

Steiner, Rudolf. *Anthroposophical Spiritual Science and Medical Therapy.* Dornach, April 12, 1921.

Steiner, Rudolf. *Art in the Light of Mystery Wisdom*, Rudolf Steiner Press, New York, 1997.

Steiner, Rudolf. *Course for Young Doctors*, Mercury Press, New York, 1997.

Steiner, Rudolf. *The Human Heart.* Dornach, May 26, 1922, GA 212.

Steiner, Rudolf. *Foundations of Esotericism*. Berlin, September 30, 1905.

Steiner, Rudolf. *From Crystals to Crocodiles*. Rudolf Steiner Press, Forrest Row, 2002.

Steiner, Rudolf. *Inner Experience of Evolution*. Steiner Books, Great Barrington, 2009.

Steiner, Rudolf. *Nature Spirits, Selected Lectures*. Rudolf Steiner Press, Forrest Row, 2007.

Steiner, Rudolf. *Rudolf Steiner on Astronomy*. Translated by Rick Mansell. The Rudolf Steiner Research Foundation, Redondo Beach, California, 1989.

Steiner, Rudolf. *Science: An Introductory Reader.* Sophia Books, Forrest Row, 2003.

Steiner, Rudolf. *Supersensible Knowledge*, Rudolf Steiner Press, New York, 1988.

Steiner, Rudolf. *The Etheric Body as a Reflection of the Universe.* Elberfeld, June 13, 1915.

Steiner, Rudolf. *The Fourth Dimension: Sacred Geometry, Alchemy, and Mathematics*. Anthroposophic Press, New York, 2001.

Steiner, Rudolf. *The Inner Experience of Evolution*, Steiner Books, New York, 2009.

Steiner, Rudolf. *The Warmth Course*, Mercury Press, New York, 1988.

Steiner, Rudolf. *Occult Science and Occult Development*. Rudolf Steiner Press, New York, 2005.

Steiner, Rudolf. *World Ether, Elemental Beings, Kingdoms of Nature.* Translated by Harold Jurgens. Mercury Press, New York, 1993.

Stoff, Sheldon and Smith Stoff, Barbara. *The Akashic Field: It Makes Every Place in the Universe Part of the Neighborhood*. Balboa Press, Bloomington, IN, 2011.

Thomas, Nick. *The Battle for the Etheric Realm: Moral Technique and Etheric Technology Apocalyptic Symptoms*. Temple Lodge, London, 1995.

Thompson, Willian Irwin. *Beyond Religion: The Culture Evolution of the Sense of the Sacred from Shamanism to Post-Religious Spirituality*. Lindisfarne Books, 2013.

Thompson, Willian Irwin. *Imaginary Landscape: Making Worlds of Myth and Science*, St. Martin's Griffin, 1990.

Thompson, Willian Irwin. *Reimagination of the World: A Critique of the New Age, Science, and Popular Culture*. Doubleday, 1991.

Thompson, Willian Irwin. *The American Replacement of Nature: The Everyday Acts and Outrageous Evolution of Economic Life*. Lindisfarne Books, 1991.

Thompson, Willian Irwin. *The Time it Takes Falling Bodies to Light: Mythology, Sexuality and the Origins of Cultures*. St. Martin's Press, 1981.

Wachsmuth, Guenther. *Etheric Formative Forces in Cosmos, Earth and Man: A Path of Investigation into the World of the Living*. Anthroposophical Publishing, London, 1932, Volume I.

Wood, David W. *Novalis Notes for a Romantic Encyclopaedi*a. State University of New York Press, 2007.

Further Reading

Andreæ, Johann Valentin. *Reipublicæ Christianopolitanæ descriptio*. Argentorati: Sumptibus hæredum Lazari Zetzneri, Strasbourg, 1619.

Andreæ, Johann Valentin. *Johann Valentin Andreae's Christianopolis; an ideal state of the seventeenth century*. translated from the Latin of Johann Valentin Andreae with an historical introduction. by Felix Emil Held. The Graduate School of the University of Illinois, Urbana- Champaign, 1916.

Arnold, Edwin Sir. *The Light of Asia, or The Great Renunciation (Mahâbhinishkramana): Being the Life and Teaching Gautama, Prince of India and Founder of Buddhism (As Told in Verse by an Indian Buddhist).* Kegan Paul, Trench, Trübner & Co., London, 1879.

Avari, Burjor. *India: The Ancient Past: A History of the Indian Sub-Continent.* Routledge, New Edition, 2007.

Barnwell, John. *The Arcana of the Grail Angel: The Spiritual Science of the Holy Blood and of the Holy Grail.* Verticordia Press, Bloomfield Hills, 1999.

Barnwell, John. *The Arcana of Light on the Path: The Star Wisdom of the Tarot and Light on the Path.* Verticordia Press, Bloomfield Hills, 1999.

Blavatsky, H. P. (Helena Petrovna). *Isis Unveiled: A Master-Key to the Mysteries of Ancient and Modern Science and Theology.* J. W. Bouton. New York, 1878.

Blavatsky, H. P. (Helena Petrovna). *The Key to Theosophy: Being a Clear Exposition, in the Form of Question and Answer, of the Ethics, Science and Philosophy for the Study of Which the Theosophical Society Has Been Founded.* The Theosophical Publishing Company, Ltd. London, 1889.

Blavatsky, H. P. (Helena Petrovna). *The Secret Doctrine: The Synthesis of Science, Religion and Philosophy.* The Theosophical Publishing Company, Ltd. London, 1888.

Blavatsky, H. P. (Helena Petrovna). *The Voice of Silence: Being Extracts from the Book of the Golden Precepts.* Theosophical University Press, 1992.

Bockemuhl, Jochen. *Toward a Phenomenology of the Etheric World: Investigations into the Life of Nature and Man.* Anthroposophic Press, Spring Valley, N. Y., 1977.

Bova, Ben. *The Fourth State of Matter: Plasma Dynamics and Tomorrow's Technology*, 1971.

Campanella, Tommaso. *The City of the Sun*. The ProjectGutenberg Ebook, David Widger, 2013.

Colum, Padriac. *Orpheus: Myths of the World*. Floris Books. Colum, Padriac. The Children's Homer. MacMillan Co., 1946.

Colum, Padriac. *The Tales of Ancient Egypt*. Henry Walck Incorporated, New York,1968.

Crawford, John Martin. *The Kalevala: The Epic Poem of Finland*. John B. Alden, New York, 1888.

Gebser, Jean. *The Ever Present Origin*. Ohio University Press, 1991.

Green, Roger Lancelyn & Heather Copley. *Tales of Ancient Egypt*. Puffin Books, New York, 1980.

Harrison, C. G. *The Transcendental Universe; Six Lectures on Occult Science, Theosophy, and the Catholic Faith*. George Redway, London 1893.

Harrison, C. G. *The Transcendental Universe; Six Lectures on Occult Science, Theosophy, and the Catholic Faith*. Delivered Before the Berean Society, edited with an introduction by Christopher Bamford. Lindesfarne Press, Hudson, 1993.

Hamilton, Edith. *Mythology*. Little Brown And Co., Boston, 1942.

Harrer, Dorothy. *Chapters from Ancient History*. Waldorf Publications, Chatham, 2016.

Hazeltine, Alice Isabel. *Hero Tales from Many Lands*. Abingdon Press, New York, 1961.

Heidel, Alexander. *The Babylonian Genesis: The Story of Creation*. University of Chicago Press, Chicago, 1942.

Hiebel, Frederick. *The Gospel of Hellas*. Anthroposophic Press, New York, 1949.

Jocelyn, Beredene. *Citizens of the Cosmos: Life's Unfolding from Conception through Death to Rebirth*. Continuum, New York, 1981.

Kovacs, Charles. *Ancient Mythologies and History*. Resource Books, Scotland, 1991.

Kovacs, Charles. *Greek Mythology and History*. Resource Books, Scotland, 1991.

Sister Nivedita (Margaret E. Noble) & Coomaraswamy, Ananda K.. *Myths of the Hindus and Buddhists*. Henry Holt, New York 1914.

Steiner, Rudolf. *Ancient Myths: Their Meaning and Connection with Evolution*. Steiner Book Center, 1971.

Steiner, Rudolf. *Christ and the Spiritual World: The Search for the Holy Grail*. Rudolf Steiner Press, London, 1963.

Steiner, Rudolf. *Foundations of Esotericism*. Rudolf Steiner Press, London, 1983.

Steiner, Rudolf. *Isis Mary Sophia: Her Mission and Ours*. Steiner Books, 2003.

Steiner, Rudolf. *Man as a Being of Sense and Perception*. Steiner Book Center, Vancouver, 1981.

Steiner, Rudolf. *Man as Symphony of the Creative Word*. Rudolf Steiner Publishing, London, 1978.

Steiner, Rudolf. *Occult Science*. Anthroposophic Press, NY, 1972.

Steiner, Rudolf. *Rosicrucian Esotericism*. Anthroposophic Press, NY, 1978.

Steiner, Rudolf. *Rosicrucian Wisdom: An Introduction*. Rudolf Steiner Press, London, 2000. GA 425

Steiner, Rudolf. *The Bridge between Universal Spirituality and the Physical Constitution of Man*. Anthroposophic Press, NY, 1958.

Steiner, Rudolf. *The Evolution of Consciousness*. Rudolf Steiner Press, London, 1926.

Steiner, Rudolf. *The Goddess from Natura to the Divine Sophia*. Sophia Books, 2001.

Steiner, Rudolf. *The Holy Grail: from the Works of Rudolf Steiner.* Compiled by Steven Roboz. Steiner Book Center, North Vancouver, 1984.

Steiner, Rudolf. *The Influence of Spiritual Beings Upon Man.* Anthroposophic Press, NY, 1971.

Steiner, Rudolf. *The Reappearance of Christ in the Etheric.* Anthroposophic Press, NY, 1983.

Steiner, Rudolf. *The Risen Christ and the Etheric Christ.* Rudolf Steiner Press, London, 1969.

Steiner, Rudolf. *The Search for the New Isis the Divine Sophia.* Mercury Press, N.Y., 1983.

Steiner, Rudolf. *The Spiritual Hierarchies and the Physical World.* Anthroposophic Press, N.Y., 1996.

Steiner, Rudolf. *The Tree of Life and the Tree of Knowledge.* Mercury Press, NY, 2006.

Steiner, Rudolf. *The True Nature of the Second Coming.* Rudolf Steiner Press, London, 1971.

Steiner, Rudolf. *Theosophy.* Anthroposophic Press. New York, 1986.

Steiner, Rudolf. *Wonders of the World, Ordeals of the Soul, Revelations of the Spirit.* Rudolf Steiner Press, London, 1963.

Steiner, Rudolf. *World History in Light of Anthroposophy.* Rudolf Steiner Press, London, 1977.

Tappan, Eva March. *The Story of the Greek People.* Houghton Mifflin Co., Boston 1908.

van Bemmelen, D. J. *Zarathustra: The First Prophet of Christ,* 2 Vols. Uitgeverij Vrij Geestesleven, The Netherlands, 1968.

Watson, Jane Werner (Vālmīki). *Rama of the Golden Age: An Epic of India.* Garrard Pub., Champaign.

About the Author, Dr. Douglas Gabriel

Dr. Gabriel is a retired superintendent of schools and professor of education who has worked with schools and organizations throughout the world. He has authored many books ranging from teacher training manuals to philosophical/spiritual works on the nature of the divine feminine. He was a Waldorf class teacher and administrator at the Detroit Waldorf School and taught courses at Mercy College, the University of Detroit, and Wayne State University for decades. He then became the HeadMaster of a Waldorf School in Hawaii and taught at the University of Hawaii, Hilo. He was a leader in the development of charter schools in Michigan and helped found the first Waldorf School in the Detroit Public School system and the first charter Waldorf School in Michigan.

Gabriel received his first degree in religious formation at the same time as an associate degree in computer science in 1972. This odd mixture of technology and religion continued throughout his life. He was drafted into and served in the Army Security Agency (NSA) where he was a cryptologist and systems analyst in signal intelligence, earning him a degree in signal broadcasting. After military service, he entered the Catholic Church again as a Trappist monk and later as a Jesuit priest where he earned PhD's in philosophy and comparative religion, and a Doctor of Divinity. He came to Detroit and earned a BA in anthroposophical studies and history and a MA in school administration. Gabriel left the priesthood and became a Waldorf class teacher and administrator in Detroit and later in Hilo, Hawaii.

Douglas has been a sought-after lecturer and consultant to schools and businesses throughout the world and in 1982 he founded the Waldorf Educational Foundation that provides funding for the publication of educational books. He has raised a great deal of money for Waldorf schools and institutions that continue to develop the teachings of Dr. Rudolf Steiner. Douglas is now retired but continues to write a variety of books including a novel and a science fiction thriller. He has four children, who keep him busy and active and a wife who is always striving towards the spirit through creating an "art of life." She is the author of the Gospel of Sophia trilogy.

The Gabriels' articles, blogs, and videos can currently be found at: OurSpirit.com, Neoanthroposphy.com, GospelofSophia.com, EternalCurriculum.com

Made in United States
North Haven, CT
04 November 2026

21766939R00178